BEHIND THE BADGE

BEHIND THE BADGE

— ★ ★ ★ —

ANSWERING THE CALL TO SERVE ON AMERICA'S HOMEFRONT

JOHNNY JOEY JONES

BEHIND THE BADGE. Copyright © 2025 by Fox News Network LLC. All rights reserved. Printed in the United States of America. No part of this book may be used or reproduced in any manner whatsoever without written permission except in the case of brief quotations embodied in critical articles and reviews. For information, address HarperCollins Publishers, 195 Broadway, New York, NY 10007.

HarperCollins books may be purchased for educational, business, or sales promotional use. For information, please email the Special Markets Department at SPsales@harpercollins.com.

Fox News Books imprint and logo are trademarks of Fox News Network LLC.

Image credits: pp. ii–iii © ecrow/stock.adobe.com; pp. iv–v © thanarak/stock.adobe.com; p. vi © kinomaster/stock.adobe.com; p. 9 © ME Image/stock.adobe.com; p. 10 © supamotion/stock.adobe.com; p. 87 © Dmitry_dmg/stock.adobe.com; p. 88 © pimonpim/stock.adobe.com; p. 171 © bibin/stock.adobe.com; p. 172 © Jonathan Stutz/stock.adobe.com

Endpaper credits: police officer © ecrow/stock.adobe.com; firefighter © Joyce/stock.adobe.com

FIRST EDITION

Designed by Nancy Singer

Library of Congress Cataloging-in-Publication Data has been applied for.

ISBN 978-0-06-343210-9

25 26 27 28 29 LBC 5 4 3 2 1

My inspiration for this book and to tell these stories came after years of sitting by a campfire and around a dinner table with the first responders in my family and close circle of friends. This book is dedicated to all the first responders throughout the United States, not just in big cities but in every small town and community where their "battleground" is—where they grew up and where they live.

We also dedicate this book to the surviving families of those who have paid the ultimate sacrifice performing their heroic duty.

CONTENTS
— ★ ★ ★ —

INTRODUCTION — 1

PART 1: ANSWERING THE CALL — 9

Ordinary Guy, Extraordinary Circumstances
CLAY HEADRICK, FIREFIGHTER — 11

The Choice to Serve
KEITH DEMPSEY, FIREFIGHTER — 35

Police of the Woods and Water
JEREMY JUDD, GAME WARDEN — 58

PART 2: ENDURING THE TRAUMA — 87

Family of Service
KATELYN KOTFILA, DEPUTY — 89

The Confidence of Camaraderie
TOMMY WEHRLE, SWAT SNIPER — 111

The Catch-22 of Police Work
JUSTIN HEFLIN, SERGEANT — 137

PART 3: DUTY OF DISCERNMENT 171

The Watchdog
VINCENT VARGAS, BORDER PATROL 173

Do Unto Others
STEVE HENNIGAN, LAPD 197

America's Sheriff
MARK LAMB 220

ACKNOWLEDGMENTS 245

INTRODUCTION

Back the Blue! A call to action I often see on social media or hear from guests on air. You see it on bumper stickers or as a tagline in a post. A simple but seemingly necessary proclamation of support for law enforcement officers. Of course, the reason for that motto is the existence of an opposing sentiment in very different circles, also expressed by the motto "defund the police" or, worse, ACAB, all cops are bastards. The battle lines of the debate about who and how we protect our communities have seemingly been drawn for us by two very different and very passionate groups of people with a message to . . . yell.

So then, where does that leave the rest of us? What should regular, law-abiding Americans think of the passionate pleas for more fair policing, or the cries of anguish from families who've lost their heroes on his or her final watch? For me, it's a relatively easy choice, I back the men and women who put their lives on the line to protect my family from the mishaps and evil acts that plague anyone living in such a free country. But also, I can understand that a mother seeing her teenage son arrested again for what she feels to be a minor offense or a crime he might've been pressured into might wonder if the officers arresting him are going looking for him as low-hanging fruit. I don't want any American judged for his or her race or appearance. I also know that in dangerous situations, it's hard to discern who's just fallen in with a bad crowd, or who's the bad crowd themselves. Would you or I, if we were in that police officer's shoes, be able to make snap judgments with perfect accuracy?

For example, imagine from any number of movies or rap songs that a young man is dressed a certain way—pants sagging, speaking with a certain slang, and demonstrating blatant disrespect for authority. Sometimes he's displaying a red or blue bandanna. Do those things make him a criminal? Is indulging in what is called urban counterculture a crime? Not at all. But, if nine out of ten of the young men who fit this nuanced description and who police encounter in the act of or having just committed a crime, how is it as black and white as racism that they treat the one that isn't as if he probably is? I'm not suggesting we don't have the freedom in this country to dress and speak however we want. I vehemently defend that freedom. But to throw a blanket over what has been proven to be an involuntary human conditioning of "if this, then that" and assign hate and racism as motives is to remove the humanity of police officers altogether.

To give an example that represents my own upbringing: Every redneck in a pickup truck with Hank Williams blaring out of his radio late on a Saturday night is a pretty sure bet to be a drunk driver, but if driving a pickup truck, wearing a mullet, and blaring the radio isn't breaking the law, should an officer treat him as such? I don't blame the officer that does, they pulled me over often when I was that loud teenager in a lifted truck. But thankfully I was never the drunk driver they expected.

But not everyone sees it that way. One of the defining moments for this national trend of young and vocal people embracing an outwardly antipolice narrative was in 2016 with the media firestorm that surrounded Colin Kaepernick, who was at the time the starting quarterback for the San Francisco 49ers. He started sitting on a bench while his teammates stood with their hands over their hearts for the singing of our national anthem. My reaction, like that of many veterans, was visceral. I was appalled and pissed off. *Who the hell does this millionaire who does nothing more than play a game for entertainment think he is anyway?* I thought.

After all, I lost pounds of flesh for that flag and anthem. I've stood graveside on prosthetic legs as my brothers were buried six feet underground draped in that flag as that anthem played in the background. I was pretty resolute in my condemnation of Colin's sitting. But then, as fate would have it, I read an open letter from former Army Green Beret and Seattle Seahawk Nate Boyer. The letter, addressed to Colin, was published in the *Army Times*. In it, Nate explained why he has so much pride in our country; he acknowledged that he believed racism was still a problem and even wrote about his own experiences in Africa that inspired him to join the Army special forces. I didn't really agree with the abundance of concessions he made to Colin, but what drew my curiosity was how Nate ended the letter, "I look forward to the day you are inspired to once again stand during our national anthem. I'll be standing right there next to you. Keep on trying . . . De Opresso Liber."

Colin responded by inviting Nate to stand next to him as he sat at the next game so that Colin could show his gripe wasn't with the military or veterans. Nate accepted Colin's invitation, but only if Colin would meet with him first. In that one-on-one meeting, Nate asked Colin to kneel with good posture rather than sitting lazily. Not unlike when a player is injured on the field. Nate explained, kneeling itself might be a sign of respect while acknowledging a problem. Colin obliged. However, the conversations with Nate weren't reflected by Colin to the media. So instead of a national discussion about a nuanced issue, we got "The Kneeling Controversy."

Nate used grace, empathy, and humility to counter what was nothing less than a slap in the face to many of us. All this was done in the name of protesting police officers and the notion that somehow police officers are inherently racist in their duties. So having befriended Nate some time earlier, and inspired by his courage and action, I reached out to him and told him I would be willing to talk to Colin as well. I did care about genuine grievances, but also knew I might be able to help get folks back on their feet and respecting our country. We had a call with Nate, myself, and Colin. After the call, I felt good, like I was helping push this issue in a positive direction. But then almost immediately afterward, Colin showed up to practice with socks depicting cops as pigs, he wore a T-shirt glamorizing communist terrorists and used my and Nate's names to justify his position rather than reach out to opposing views. He suckered us. That's my experience with the "defund the police" folks. They were attention-seeking crybabies with no plan of action other than to prey upon the guilty conscience of woke corporations and the bad nature of spineless politicians. They were never trying to connect with people; they were trying to sabotage policing in departments across the country and enrich themselves.

For a few years at least, they did just that. Our communities grew less safe; our police officers were indicted on oftentimes racially motivated and bogus charges. By 2020, riots and local insurrections were both common and *justified* in liberal, Democrat-run cities across the country. That's what "defund the police" got us. Pain and injustice.

Acutely felt among the pain came instances like the 2016 ambush and killing of five Dallas police officers at what was supposed to be a nonviolent social justice march. A string of execution-style murders of police officers followed across the country. And, more recently, the 2024 murder of NYPD Officer Jonathan Diller, who was shot and killed conducting a traffic stop. All of these exacerbated the tensions and real pain felt by those on the other side of this issue. It seems for every hate-filled protester there was also a family mourning their slain hero. Perhaps not all directly connected, but heartbroken just the same.

So one side countered the hateful narrative of the other; they put stickers on their trucks depicting American flags with a thin blue line across the center. They coined "Back the Blue" and proudly exclaimed their support for the men and women who protect and serve us all every day. That's certainly the side I'm on. I do back the blue, and red, and green, and all domestic heroes whose occupation is to put their own lives in danger each day to keep us safe. Although I must admit, I still get angry when a state trooper impedes my hurry on the interstate with a speeding ticket.

So, where do these battle lines leave us as a society? More recent events, including the 2024 presidential election, have shown that our politicians just might be hearing the cries of their constituents who feel unsafe: voters who overwhelmingly decided more policing is needed. But the racial tensions fueling the antipolice narrative are still alive and well. Some even on the right side of the political spectrum see local police officers as nothing less than tyrannical government agents who only exist to infringe upon their rights. Even a trend of self-proclaimed "sovereign citizens" attempting to catch police officers mishandling justice has taken hold on social media. So then we must ask: How can we reconcile the relationship between those who serve us and the fear we have for the authority they hold in doing so?

Something I've realized in hearing from the first responders in this book is that the work they do is more complicated, unusual, and dangerous than I could have ever imagined. Much like how movies incorrectly portray urban teens in a highly negative light, they also gloss over the true nature of being a police officer or fireman. First responders contribute to our peace in ways I'd never imagined, but in doing so sacrifice their own. They bear a constant drip-drip-drip of trauma that leaves me amazed at their grit and perseverance. What sacrifices do they make to do their job? How does it affect their lives, relationships, mind, and body? Perhaps the first real step to getting beyond these opposing mottoes is simply understanding who these people are behind the badge they wear.

And not just those with blue lights and a gun. But those that save our burning homes, those that find our loved ones when they've gone missing, and those that protect our borders and even save those who try to cross them illegally.

I have two amazing friends I get to call brothers . . . in law, Keith Dempsey and Terry Mills. They both are currently serving as chief officers within their respective departments. They serve in neighboring departments; one where I live and the other in the town next door, where I grew up. They have spent their entire careers trying to save lives in our community. When they take their families to dinner, they drive through their battlefields: intersections where they have responded to a terrible car wreck, or apartment buildings where domestic violence and drug use has them there weekly applying first aid and resuscitation. They see the absolute worst that we can do to each other as human beings, yet they still show up to work every day. They are ever engulfed, not just in flames, but also in the pain and sorrow that plagues any community.

I've seen firsthand just how much weight a first responder carries even weeks after a tough call, and it has given me a whole new appreciation for and curiosity about the work they do.

The Tuesday before Thanksgiving in 2019 my maternal grandmother passed away. The funeral was the Friday after, so we waited a week to celebrate the holiday. The following Friday my sister, who was single at the time, met my parents at my house, a few hours south, and stayed the weekend with my family to give thanks and enjoy one another. My dad, who had been in poor health for a few years, seemed even worse. My parents arrived late that Friday; they are renowned for their lack of time awareness, so an evening of lovingly ribbing them ensued. It was fun; my dad and mom took jabs at one another laughing and poking. It was a good night.

The next day we invited our neighbors, who had become like family, over for a Thanksgiving meal. I was outside shooting guns most of the day and my dad took my son on a "walking adventure" in the woods behind my house, where they found a big snake in the cab of an old, abandoned truck. That night we all stayed up playing card games around the dining table, but my dad went to bed early. Once we all got ready to call it a night, I rolled my wheelchair back to the guest room, cracked the door, and told my dad good night. At around 5 a.m. the next morning I awoke to a bloodcurdling crying scream, my mom's voice. I jumped in my wheelchair and rushed back to their room. In there I found my dad sprawled out on his back on the floor; he was lifeless. My sister was already performing CPR. Knowing that once you start you don't stop, I jumped out of my wheelchair and relieved my sister doing chest compressions and she started breathing in his mouth. Meg, my wife, called 911 and in what seemed like just a couple minutes firemen showed up with their medical bags and took over working on him. The memory of that night is forever burned in my head.

I rolled into a laundry room that had a window facing the front of my house. I sat there sobbing, praying, hoping to hear good news as they took him out to the ambulance. My mom and sister rode to the hospital with him and I arrived an hour later with my family. Eventually they'd kept his body alive, but his brain was gone, and we made the tough decision to remove the machines and let him rest in peace. I was in there alone with him holding his hand as he took his last breath. It was traumatic and sad and terrible. But I also had the opportunity to tell him goodbye. To say things I needed to say to a dad I'd too often butted heads with, but also greatly admired. A dad who was better at protecting and providing than consoling or listening.

A few weeks later, I was approached by a man in a uniform outside a local business as I got in my truck. The look on his face told me all I needed to know before he even spoke. He identified himself as a commander or captain or something of that nature with the fire department, and said he was there that night as the officer in charge. His face showed both remorse and hope as he asked, "How did it turn out with your dad?"

When I told him he passed away the next day the man hung his head and simply said, "I'm sorry we couldn't save him." I smiled and thanked him for giving me a chance to say goodbye. We shook hands and parted ways. What we shared that night in my house was one of the most difficult moments of my entire life, but it was just another shift for him. As far as I know, he had other similar calls that same night. But he remembered, he cared, he wore the burden on his face and the sorrow in his eyes.

How much of that kind of trauma can one man or woman carry, and for how long? Perhaps that's the question we should consider as civilians in a country so free of danger because we are so blessed with first responders who sign up for the job and responsibility of trying to save us from it. Perhaps rather than fighting over whether to support or attack first responders, we should be reflecting on having gratitude for the job they do. Starting there will give us a whole new appreciation for what goes into the work of first responders. That's the question I sought out to answer. And I didn't have to look much further than some family and dear friends to get the answers. In the following chapters of this book, I interview nine first responders who've all had an impact on my life. Most of them lifelong friends, some family, and all of them heroes to me. In this raw, unfiltered look into the duty they fulfill and the burden they carry, I get to the heart of what it means and what it costs to bear the burden of selfless service in your own community.

PART 1
ANSWERING THE CALL

I look at it this way. I'm just an ordinary guy who has been well trained and finds himself in extraordinary situations. I try to tell everybody, especially those looking to become a fireman, who might listen that the things that we do in the fire department, if you had been trained as long as we have to do those things, you could do them too.

—Clay Headrick

I can't really explain it, but from the time I was a preschooler, I wanted to be a cop. . . . Unlike a lot of young people, I never let go of my early answer to the "What do you want to do when you grow up" question.

—Steve Hennigan

I was hooked. I was blown away by what the life of a trooper was like. We were chasing guys down the highway at 120 miles per hour and making stops and taking guys to jail for drug possession. And we got paid to do that? And I could be on a bomb squad? I knew that there was more to it than that, but comparing that adrenaline rush to the tasks I completed as an electrician? There was no competition.

—Justin Heflin

Ordinary Guy, Extraordinary Circumstances

CLAY HEADRICK
★ Firefighter ★

"At some point, this job is going to hit home. Something is going to physically, spiritually, or emotionally break you."

THINGS YOU CAN'T FORGET

Something happened pretty early in my firefighting career that just about broke me. I was one of those young guys who are full of piss and vinegar and fire. I really wanted to learn. I wanted to be in on the action. If something was going on, I wanted to be a part of it. One day we ran a medical call, a very traumatic one. We showed up on-scene and learned that a child had been run over by a truck. It happened right in front of his family. Mom and Dad and the rest of the family just lost it. I could understand why. I really don't know if it was the trauma of seeing that child in the road that got me, or if it was the emotional aspect of what the family was experiencing.

That call hit me really hard. Maybe it was because I had very young kids myself at home at that time. I was a brand-new fireman, and I just lost it. I was totally done. Luckily, I had a pretty good company officer who was there with me. And he could tell I was off in left field.

I told him, "I can't do this. I can't do it. If that's a problem, I'll lay my badge on the truck bumper, and I'll walk back to the fire department, and I'll be done."

He looked at me and, pointing to a spot away from the scene, said to just go over there and sit down. Hang out over there. Chill out. That was pretty much what I did until the call was done and we went back to the fire station.

★★★★★★★★★★★★★★★★★★★★★★★★★★★★★★★

> Clay Headrick is someone I've known since I was two or three years old. To hear him tell it, I walked up to him, barely out of diapers, climbed into his lap, looked at him, and proceeded to punch him squarely in the nose. I don't know why that matters, but it cracks me up and is a pretty good way to explain just how much like family he is.

Growing up, he was my uncle Jeff's best friend, and they had a dirt-track race car they raced on weekends. Jeff owned the car and was the driver. Clay's role was to fabricate parts and adjust the car for different track conditions. Together, with little help and less money, they won a lot of races in the small towns of northwest Georgia and southeast Tennessee.

My first book, *Unbroken Bonds of Battle*, featured nine military veterans and one Gold Star widow, all of whom I have known for several years. They weren't just veterans I met; they were childhood best friends and brothers. I stay connected with them through a few group texts and some annual hunting trips I put together for us.

But along with the veterans who attend those hunting trips are a couple of firemen I've known even longer. Clay is one. Keith Dempsey, the subject of the next chapter, is the other. Keith, along with being my brother-in-law, is one of Clay's best friends and a fellow fireman at their department.

Getting Keith and Clay involved in these hunting trips was an idea that came to me after hearing Keith talk about being a fireman at family gatherings. He, like many first responders, doesn't talk to the rest of our family about his job, but having a combat veteran like myself join the family meant he made an exception for me. We've grown close and shared many stories with each other.

Unlike many, Clay came into the fire service without a direct influence. He didn't do the explorer program, and he didn't have a ton of preparation for what the job would really be.

But he had a desire to help. He had a passion for learning and felt a calling that the fire service would fulfill. Even still, I know full well what the phrase "baptism by fire" means for your first moments in combat, and Clay experienced something

> tragically similar that day early in his career. Statistically, the vast majority of calls that firefighters respond to aren't fires. Unfortunately, medical emergencies including tragic accidents can be a much harder pill to swallow. The fact that anyone can do that job and see what they see is quite extraordinary.

★★★★★★★★★★★★★★★★★★★★★★★★★★★★★★

AN ORDINARY GUY IN EXTRAORDINARY SITUATIONS

I look at it this way: I'm just an ordinary guy who has been well trained and finds himself in extraordinary situations. I try to tell everybody, especially those looking to become a fireman, that the things we do in the fire department, if you had been trained as long as we have, you could do those things too. Some folks kind of balk at that and say, "No way." But I remind them that in the beginning, I wasn't sure if I could do it either.

Part of what we do in evaluating and training candidates is to expose them to elements of the extraordinary. We also evaluate them for weaknesses, like if they have trouble breathing when wearing a mask or functioning in a darkened environment. Sometimes they have issues with squeamishness in dealing with bodily fluids and other things they might encounter during a medical call. You just never know what might pop up. But once it does, our goal is to help them work through it and not use that as a reason to disqualify them. We talk them through it, expose them to techniques to help overcome those weaknesses, and assure them, as my company officer did with me, that "we will get you through this." And most of the time, with training and repetition, you show people that "Hey, I can function in this environment, and I can be a benefit to other people."

So the whole "hero" thing? I don't know about all that. I'm just an average guy doing a job that he loves.

LIFE ISN'T HANDED TO YOU

My dad, Cecil, was a mechanic by trade. I come from a very long line of mechanics. He quit school when he was very young—a teenager. His father had died, and he needed to go to work. He was a very self-educated person and is one of the few people I have ever seen who could pick up a book and read how to do something and then just go do it. That amazed me.

My mom is Brenda, and I've got a brother who's six and a half years younger than me. His name is Jon. I had a pretty much average childhood. Mom and Dad separated when I was around thirteen. I lived with my dad until I was about sixteen, and then, well, let me put it this way, we kind of butted heads. I was a stubborn teenager and he was a hardheaded person himself.

I ended up moving back in with my mom. I love her to death, and she's always been there for me. But we're very much opposites with different values, in how we act, and there's nothing wrong with that. We're just two different people.

My mom has always been a very professional person, and money and status matter to her in a way that they don't to me. Don't get me wrong, it would be great to have a lot of money, but to me, being with good people and doing good things has always been more important. Not that my mom isn't a good person or did bad things, but it's just a question of perspective and priorities, I guess. I don't think I'm right and she's wrong; we're just different.

But with my dad, he kind of felt like the world had done him wrong in some way. The world owed him something. That's kind of what we butted heads about, him being negative like that. If you want something in this life, you're going to have to work for it, and go out and get it. The thing was, he's real smart, real hardworking, really good with his hands, and got to work building custom hot rods for a local guy. Maybe dropping out of school at a young age was what set him back in his mind. He didn't want to take risks, so he didn't pursue some opportunities. Too much risk. But you can't navigate this world thinking everybody owes you or that everyone's always trying to get even with you.

I've spoken with some counselors about things because I've had some issues over the years, and they always want to go back to childhood. It seems like that's a cookie-cutter approach, and I get aggravated with that. I understand using it with somebody who was abused or neglected as a kid. But I wasn't. Sometimes you just have parents who love you and have normal issues, but that shouldn't define your life. Everybody is their own person. I believe that I'm one hundred percent responsible for every part of my life. I don't want to make it sound like I'm throwing my dad under the bus.

Still, you need people around you. That's why it was great timing when the Jones family came along. When I was sixteen I met Jeff, Joey's uncle, on a Saturday night at our local racetrack. He asked me if I was interested in helping him with a race car. And that's how my life with the Jones family began. They came along at the right time. For the next ten years, I was pretty much a Jones family fixture; they included me in pretty much everything and anything that went on with them.

★★★★★★★★★★★★★★★★★★★★★★★★★★★★★★★★

Over the years, I've realized that there is a lot to be learned about the kind of person Clay is when you think about the way I got to know him. My uncle Jeff was a car-racing prodigy of sorts. He's very outgoing and nice; he's funny by nature and extremely kindhearted. People just want to be around him. But when he gets behind the wheel of a race car, he's a fierce competitor, and that attracts a lot of attention and adulation.

Clay's role with Jeff and his car was just as important, but wasn't as front-facing. Clay is very smart and engineering-minded.

The undercarriage of a dirt-track race car is essentially a network of bars and pivot points that move independently to turn the car around the tight corners of an oval track as fast and efficiently as possible. To do this, you have to understand principles like weight distribution, centripetal force and centrifugal force, and other complexities of motion. Clay didn't have a formal education on these things, but he has an eagerness to learn and a mind that could "figure it out."

He worked tirelessly to set the car up so Jeff could "drive the wheels off of it." They couldn't afford the big-horsepower engines and fancy equipment other teams had, but they knew they could outsmart them and outdrive them. As extraneous as that may seem to Clay's decision to become a fireman, it's actually quite telling. If he was willing to work that hard for a trophy that Jeff got to keep, it's in his DNA to work that hard to save a stranger's life or property. He doesn't need to be called a hero; he just wants to do the hero's work. But like anyone who risks his or her own life, and does so to selflessly help others, it eventually takes its toll and comes at a price.

★★★★★★★★★★★★★★★★★★★★★★★★★★★★★★★★★★★

DEALING WITH THE FALLOUT

I've been a fireman and a paramedic for twenty-eight years. Somewhere along the line, I think something broke in me. Either I ran the wrong call or saw too much . . . It got to the point where I knew I needed help. I went to speak with a counselor, a couple of different ones in fact, over the years. I got aggravated by their approach. Trying to blame everything on something that happened in my childhood. Like I said, I'm one hundred percent responsible for my life, including what I've been subjected to and what I allow to bother me and what I don't. Still, I had to take that step to get help.

When that happened with the child being hit by the truck, it really hit me hard. And after everything was over, we went back to the fire station. You're physically away from the scene but all kinds of things were still on my mind, of course.

I was married and called my wife. I try not to go into details with her or other family members about what you see and what you do. But I think she knew I was really upset just by the tone of my voice. We didn't talk, just for a few seconds when I asked her to put the kids on the phone. I told her that I just needed to hear their voices and know that they were okay. After I talked to them, I sat on the back of one of the trucks and cried for an hour. I really almost quit the fire department after that. I didn't think I could mentally or physically get past that point. Looking back on it, I think that's when I first got broken.

While I sat there and wondered, feeling scared to death, the alarm went off again, and at first, I thought, *Is this going to hurt even worse than the last call?* I also understood that sometimes you show up on a call, and it isn't all bad and you can make a difference in the lives of the people. Luckily, that's what happened on that next call. That helped me get back on my feet. And for twenty-eight years, that's how it's been. I've struggled with that tension between getting hurt by what I do and feeling a sense of achievement, satisfaction, or what-have-you. But that continuation over year after year after year, of seeing people hurt and dying, going through loss and tragedy, it just kind of builds on those bad thoughts.

When I do open up to people, to family, friends, and colleagues most of them, if they haven't experienced what I have or have the same issues, they tell me not to think about it, don't worry about it. I wish there was a switch I could flip and turn those thoughts off and put those bad days behind me. But it doesn't work like that. It's lying in bed for an hour every night trying to go to sleep. Then, once you do, you have nightmares about it half the night. Then you wake up the next morning, and it's the first thing you think of. It just doesn't turn off. And my nightmares and anxiety kept getting worse and worse until I was almost constantly asking myself if I could continue to take this.

Luckily, I had a good internal medicine doc as my personal physician, and whom I consider a friend. I pretty much went to his office one day and broke down right there. I told him that if this is all life has to offer, and if this is all I've got, I don't want to keep living like this. I can't keep living like this. He was smart. He was aggressive, and he helped me get treatment and some medication.

That got me going, kind of got me squared away, and through years of dealing with it, processing it, and talking about it, it's gotten manageable. I'm not going to tell you that I don't have it anymore, because I do. It's something I'll probably deal with for the rest of my life. So, I have highs and lows, just like anybody else. But sometimes, when I have a low, I have a really bad low.

PREPARING OTHER PEOPLE FOR THE TOUGH STUFF

One of the ways I can use my difficult experiences is to help people new to the profession. With me being a paramedic, I'm also one of our medical instructors for the fire department. I also used to do some teaching at the local hospital. I make people aware early on that they may think they are bulletproof, but they're not. And that's okay. You may think you're the toughest alpha male or the toughest alpha female, but at some point, this job is going to hit home. Something is going to physically, spiritually, or emotionally break you, and you can't be afraid to reach out for help.

In today's society, one of our largest killers in public safety jobs is suicide. I think that's because a lot of us think of ourselves as that alpha type, and when we need help, we don't want to ask for it because we'll seem weak. People don't reach out for help until it's too late. Since I've started talking about how fallible we are, how even the strongest of us can get laid low, and discussed my own struggles, I've seen some real changes. The number of people who have reached out to me to talk about their personal problems or struggles is amazing. I can't say that I've helped them with those problems, but I've given them guidance about what they could do to get help.

I also made it known I would much rather they call me in the middle of the night. If I have to get in my truck and come to your house or wherever you're at and get you or sit and talk with you, I would rather do that than go and spend a week dealing with your funeral. So, helping people to understand that has been a big way for me to help myself, because it allows me to tell my story and basically open up myself to somebody else. So that has helped me a lot.

★★★★★★★★★★★★★★★★★★★★★★★★★★★★★★★★

Clay hits on something I've learned to be incredibly valuable in processing our emotions and experiences. It's not just the satisfaction and fulfillment of helping others that relieves his own anxieties when sharing his story. It's the process of talking about it himself. Traumatic stress and anxiety can build up in us like a disheveled, messy filing cabinet with a door that has been forced shut. It sits there with all this tension of compressed papers constantly pushing against the door, trying to force it open and relieve the pressure. Those are our negative experiences when we suppress them. When we finally talk about them openly and honestly, something amazing happens. That filing cabinet slowly opens, and the messy papers start to become organized and formed. And eventually we can shut that cabinet back with ease and without any tension of compressed messes. That's why simply talking about our trauma is not just processing it for ourselves but also helpful for others to hear it as well.

★★★★★★★★★★★★★★★★★★★★★★★★★★★★★★★★★★★★★★

TRAINING FIREFIGHTERS

The City of Dalton Fire Department has an extremely positive reputation in the state of Georgia. When you talk to people about us, most of the time we're perceived as a very aggressive department with very high standards. For example, our recruit school. When you get through the lengthy application process, you sign on for fourteen weeks of recruit training. That's actually Keith's job. He's in charge of ensuring that every recruit meets the standard we set for a Dalton firefighter. And at this point, he's also written that standard.

Recruit school is meant to train you in two things: 1) what you *think* you're getting ready to get into, and 2) what we will actually *need* you to go do. So, in those fourteen weeks, those candidates, with the exception of just a few days, will do physical training every morning. Our PT is normally anywhere from an hour to an hour and a half, followed by normally six to seven hours of classroom work each day. So you're getting physically and mentally exhausted every day.

One thing you learn in training for the fire service is that you have to push somebody to physical and mental exhaustion to find out if they have the heart to still go on and keep doing what they need to be doing. One of the things that I've noticed after doing this for nearly thirty years is that a lot of times you get these big, stud athletes who have never really struggled with anything in their life, and all of a sudden you put them in an environment and they are struggling physically.

They don't know how to deal with that because they've never struggled with this type of physicality before. Lifting weights, running, even football drills are strenuous. But doing it with an air tank in extreme heat and low visibility adds a mental element they haven't experienced. So, once you expose them to that, and they see that physically they're breaking, then you really have to teach those guys how to mentally deal with the task at hand. Also, a lot of the time, the people who have always kind of been middle of the road were physical; they could do the job but they weren't the star athletes. Everything they had to do in life, they had to struggle and put forth a lot of effort and time into. They've built their mind along with their muscles. Those people always seem to excel because through their life, they've been used to facing adversity in everything that they do. So when we put them to the test and subject them to adversity, they know how to deal with it and so they pick up, go on, and do what they need to do.

In those fourteen weeks, they will gain everything they need to become a state-certified firefighter. They will have an operations-level hazardous materials certification. They will also be certified in emergency medical rescue (EMR), which is basically entry-level, and they'll get exposed to a lot of other stuff there too. They'll also be trained up in vehicle rescues and things of that nature, and some specialized rescues. Then, normally, in the last two weeks the recruits go through what we call burn weeks. That's when they get in full gear. They've been exposed to their gear and their self-contained breathing apparatus (SCBA) pack, which is the air tank they wear on their back. They've been exposed to those throughout recruit school, but now in those last two weeks, when they come in in the morning, they will do their PT in full turnout gear with a SCBA pack on. They will be in their gear all day long, and they're going to be throwing ladders, they're going to be fighting fire, they're going to be doing search and rescue. They're going to be doing down-firefighter drills, getting each other out.

The pack is just one part of gearing up fully. A lot of times, when we come off the truck, we always tell our guys, You need to have tools in your hand. The vast majority of time, we won't use them, but if you don't have it, you can't use it. So we like for people to come off the truck with a tool in hand, whether it's an ax, a set of irons, or a hook. And that stuff gets heavy.

I know that for me personally, with everything that I carry, and my boots, my bunker pants, bunker coat, radio helmet, SCBA, face piece, a thermal-imaging camera, and a tool, I'm pushing right at 300 pounds. I weigh a little more than 220, so that's eighty pounds of gear. That doesn't sound bad, compared to what soldiers carry. The big difference is that bunker gear is like wrapping yourself up in a thermal blanket. It's heavy and thick, and that keeps you from getting burned, but it doesn't keep you cool. So we crawl into a structure, and it's 200–300 degrees on the ground. I always compare it to being a baked potato. You get hot—really hot.

But that gear is a necessity. We don't want you burned, and we don't like to get burned either. So there is no way around it: it is extremely hot beneath that gear. You're wearing eighty pounds of it. You can't see your own hand in front of your face. But you have to go to work, and you're going to exert yourself as hard as you've ever done.

There's a method to the madness, and in this case it's to be sure you can function in that environment. You see, we function as a team. Everything we do, we do as a team. So, if the guys on my truck or team have a problem, it's everybody's problem, because we either have to take care of them, save them, rescue them, or, at bare minimum, pick up their workload.

That element of the training proceeds in steps. It starts out with them crawling around in our burn building or our smoke tower. Then we'll black out their masks and see how they do in a blacked-out environment. They have to learn how to use their hands to feel and navigate blind. Then we'll remove the blacked-out mask and add smoke. So now they know, okay, well, if I wig out and I take my mask off, I'm not going to breathe anything but smoke.

So, the psychological aspect of it keeps getting turned up as you go. Once they get past the cold smoke drills, we start putting them into heat. Now they have the blacked-out mask with the addition of high heat. You have the stress of being in all the gear, and it takes a toll on you. You're mentally and physically shot at the end of the day.

It's fun to watch somebody come in on that first week. You look at somebody and you're like, *I don't know about that one*. And then, fourteen weeks later, that person is leading your class. So, you get to watch somebody transform in fourteen weeks into a firefighter that we'll be working side by side with. And we're comfortable with going to work with them.

When we bring people in, we expose them to these things. We look for weaknesses in them, but not to get them out of the program. We need to know what they need to work on so that we can help them overcome those weaknesses. For example, if they have problems feeling claustrophobic and breathing normally when wearing a mask, or they struggle while being in a blacked-out environment, or if they have issues with medical calls or whatever. We want to expose those issues as soon as possible so we can train them. We can work with them. We can talk to them. We will get them through this. The last thing we ever want to do is turn a candidate away. Most of the time, with training and repetition, you show people that, hey, I can function in this environment. I think most everybody has the capability of doing that once they've been trained to do it.

> Georgia, like all American states, has a rigorous set of standard skills that each candidate is tested on to become a certified firefighter. Under Keith's leadership, and with the help of seasoned veterans like Clay, their recruits are trained and tested even beyond the basic requirements. Unfortunately, training them for the psychological stress isn't yet standardized. Training them to deal with the trauma and stress that comes from it isn't as easy as a practical application test, it takes real experiences and people like Clay who've gone through it to recognize their struggle and help them talk through and process it before it becomes a problem.

WORKING WITHOUT A SAFETY NET

There are so many sides to this, it's ridiculous. Being a part of the team puts added pressure on you, as well as providing you with a lot of positives. How we, as individuals, perceive ourselves is important. And that creates a bit of tension sometimes. We want to have the image that we're indestructible and that we can do anything that gets put in front of us. It may take us a few minutes to get something done, but we will prevail. We will make it happen. So that's a very good attitude to have toward most things we do.

But it can also be dangerous. We have new people come in, and they want to prove to the other guys who have been there longer that they know how to do the job. They want to prove that they are a valuable member of the team. So they push themselves, and sometimes they do that until they're at their breaking point. Luckily, if you're matched up with good company officers and good people on your crew, they will watch you, watch out for each other, and watch each other's backs. If we see you starting to get in trouble, or struggle, for example if you starting to breathe too heavy, like you're potentially having medical issues, then I'm going to get you out of the fire environment.

We do that in the real environment and in the training environment, but mostly that happens during training when we're in the burn building. Yes, you can get hurt in the burn building. Yes, you can die in it. It has happened all over the world with firemen. But you know that there are instructors in there with you, and your every move is being watched in that building. You have a safety net. You know that if something happens, those guys are going to get you out.

You really don't have that same safety net on a real fire. Yes, we are there. And if something does go wrong, we will do our absolute best—even to the point of dying—to save you. But you have to understand that you don't have that full safety net. You have to understand that this is real and has real consequences. You can die. The people around you can die. So, when you mix all that together and you're in there fighting a fire, a lot of things are running through your mind.

I remember the first structure fire I ever went in on. I was so excited and scared to death at the same time. I've talked about it with other firemen that were there. Still, to this day, I can say it was probably one of the top five worst structure fires I've ever been in in my life. It was the first one that I went in, and it was really bad. I mean, it was hot. You couldn't see anything. We couldn't find the fire. We were having all kinds of issues getting the hose line to the back of the structure through the inside of the house. The original house had had an addition built on to it. The old construction was tongue-and-groove hardwood lining the inside. So it was holding a lot of heat. You can't hear very well in those environments. A lot of people hollering, trying to communicate. Years ago we didn't have the best radio system in the world, so a lot of it was mouth to mouth and trying to talk to somebody in there, you're basically having to yell at them. You've got this face piece on, and every time you breathe in and out, you sound like Darth Vader in *Star Wars*. And with everything going on, there is a lot of screaming and hollering.

We got the fire out and so luckily saved the house. I sat down in the street in front. I sat on a big metal box that we used to carry saws in and thought, *What in the hell have I got myself into?* Because it was crazy. I was ramped up on adrenaline but also scared to death. And honestly, when I went in there, I didn't know if we were going to come back out. It was just an emotional and physical overload.

When we got back to the station, we were sitting in the kitchen and I was eating something. One of our senior firefighters came in and asked me, "What do you think about your first fire?"

I opened up a bit to him: "Dude, I'm—I don't know if I'm going to be able to do this. That was . . . I didn't expect that.'"

He had been doing it long enough that he could say, "Listen, that was a pretty bad fire, okay? That's going to be close to the worst that you'll see working structure fire." And I thought, *Well shit, if I've seen the worst on my first house fire, then maybe I can do this job.*

So, I gained some confidence there, and I kind of picked up my spirits. Every time you go, you get a little bit more comfortable with it, to the degree one could call it comfort. You get a little bit more confidence in your skills and what you can do, knowing that you can function in that environment. It makes it a little easier to deal with and do your job. I always tell candidates when they come through recruit school that when somebody calls 911, that's not an emergency for us, it's a job. Their emergency is our job, and we've got to be able to do our job, and it takes time for us to mentally react calmly and deliberately.

On that first structure fire, I was definitely in emergency mode. I imagine some of my thoughts were like what a homeowner might have when their house is on fire. It was a good thing that I could be honest with that older guy and gain some perspective. Who knows what would have happened if I hadn't spoken with him and gotten the benefit of his experience?

I heard a saying a long time ago when I was very young in the fire service. It always stuck with me. It goes: In the fire service, we kill our old, and we eat our young. Meaning that most of the time, when you're coming into this profession, you're thinking that the instructors are trying to kill you. The training demands are that hard. By the time you get all the job knowledge to really be confident and competent, you don't have the same body that you did when you were in your late twenties. You don't have that stud body anymore, but you've got that knowledge. And, as a younger man, you see this and believe that none of these older guys can hang with you.

I'd look at these older firefighters and think, "Dude, your ass is old. Why don't you go on and retire? You can't do this, and you can't do that." But what I realized later on was that "old dude" brought to the table thirty years of experience, and that the things that you can't learn out of a book, they've learned out in the field, and they can deal with it. So sometimes, early on mostly, I'd go out on a call and encounter something I'd never seen in the manual—whether as a firefighter or a paramedic. I didn't know what to do. That old, salty guy who I thought didn't need to be there anymore would know what to do. Chances are it would be something simple and straightforward.

It's kind of like with a football team. Having those guys with experience to mix in with the youngsters is important. It's also important that everybody understands their role. Everybody has their strengths and weaknesses. We don't want a kicker throwing the ball. It's important to have that mix of young and old on a shift. As a veteran, you're constantly having to teach the new guys. They're all gung ho, and they're attending all the latest classes and learning about all the new cutting-edge equipment and whatnot. And that's great. And they're teaching us older guys about what they learned. And that's great. But the old guys are teaching them about how to survive in the field. So there are two sides to that coin, and that's what really makes this work.

***

> Clay understands that the vast majority of calls they respond to are medical emergencies, because there aren't as many fires as there used to be. This is, in large part, a good thing. It's a result of decades of learning from the past, building structures with safe materials and with fail-safes like sprinkler systems and smoke detectors communicating with security systems, as well as educating the population on fire prevention and safe living practices. But while the specific skill set of fighting fire may be used less often, dealing with people in medical emergencies is a growing challenge.

***

"PEOPLE ARE THE DANGER"

Any good firefighter will be able to tell you how a fire is going to progress. I can sit in a training class and look at structure after structure after structure. You can tell me what that structure is made of. You can tell me what's contained in it. You can tell me where the fire is going to originate and how it's going to progress. But you can't do that with people. People are the unknown. People are the danger because you can't reliably predict how they're going to react in a situation. You could be called out to the same house regularly, and the people will just be as happy-go-lucky and glad to see you as could be. Then, the next time you go, they've got a gun and are threatening to kill everybody there.

So, people are the danger. We talk about some substances being volatile, unstable. Well, that's very true of human beings. I started out feeling all kinds of fear and stress on that first crazy structure fire. Now reading people is really the hard part of the job. It's stressful trying to figure out these people and what they're going to do, or what they're capable of doing, because you just don't know, and there's no book that you can read or teacher who can tell you because people are all different and they're going to respond differently.

★★★★★★★★★★★★★★★★★★★★★★★★★★★★★★

> Firefighters don't work on an island. The work of emergency response requires multiple agencies and skills. They work alongside police and law enforcement when they get on-scene. They hand off emergency medical patients to EMTs when the ambulance arrives. But before any of that, they are dispatched and directed by the 911 call center. That's something Clay has personal experience with. Along with riding the ambulance as a paramedic, he worked in the call center on his days off from the fire department for a while.

★★★★★★★★★★★★★★★★★★★★★★★★★★★★★★

THE UNSUNG VALOR OF DISPATCHERS

Those people really don't get the credit they deserve. I know this about our area in particular. During the daytime, we have an influx of one to two hundred thousand people to the area to work in the various textile mills. Interstate 75 also runs through town, making it a major travel corridor. Everybody has a cell phone now, so if anything happens, they have the capability of calling it in. That's a good thing, of course, but if there is one wreck on I-75, three hundred people may call that one incident in. The dispatchers have to handle each of those calls, vet the callers, and get all the details to make sure they're not phoning in a *different* wreck. They can't assume it's a call about the previous known incident.

So, when something does happen, those offices turn into a complete madhouse. Eventually the stress of that got to me, and I couldn't do that work on top of my firefighting and paramedic shifts. I admire those who stick it out. In my mind, they're like teachers—underpaid and sometimes underappreciated.

I have to say, that was a stress I didn't need. I love to go out and do things and socialize. I used to like going to concerts and seeing people and going to racetracks and you name it. Now when I go somewhere, if there are a lot of people around, it puts me on edge. I don't know what they're going to do. It's like I'm always on the lookout for something bad to happen. I didn't sign up originally to be a firefighter thinking that I'd be dealing with people this much. There's a whole lot of public contact involved in what I do. And that's taken a toll on me. So I'm one of those people who enter a room and are always checking for exits. Most of the time I sit or stand with my back against the wall, and I try to keep people at arm's length. When I'm out, I'm in situational-awareness overload.

I've cut back on going out and doing things. I've got my inner circle. I've got Keith and Joey and the vets we hunt with, and a few other friends. We talk. If I have a problem, I can pick up the phone and let them know I need to talk. And they tell me they got me, and they do. I've got my inner circle, and then, if I'm not in my inner circle, then I still try to stay attached to other people who work in public safety. We read each other well.

One of the main things I'd suggest to people, whether you call them civilians or laypeople, is to think of us and treat us like you would a soldier. When you approach a soldier who you know has seen combat, would you ask them about the time they killed somebody?

That's the last thing that dude wants to think about or talk about. So why would you come up to a firefighter and ask them to tell you about the worst thing they've ever seen? When people do that with me, I'm thinking, *Why would you want to drum that up for me? Can't we talk about how pretty the scenery is or how fresh the air we're breathing is, or how good a time everybody seems to be having? Why bring that up since you're not ever really going to be able to understand fully what I went through?*

I'm also troubled by so-called stolen valor—when people falsely claim that they served in the military or in public safety. They'll tell you they did this, or they did that. Why? Why lie about that? Why take away from what the folks who really did those things experienced and sacrificed? The other thing is, I don't care how good a liar that person is, how much research they've done, if they came up to me and started talking about firefighting or emergency response work, it wouldn't take me but two or three minutes and I could see right through them. Firefighters, police, and other first responders sometimes have to deal with that kind of disrespect.

That's why I enjoy being at the fire station so much. We're there one-third of our time each week. But we're all the time messing with one another. Yes, sometimes we talk about things we've got going on in our minds and our lives, but a lot of the time it's joking; it's cutting up. We play pranks on one another. We laugh at one another if we do something stupid. So a lot of good times happen there. You build up a bond with the guys. Maybe you cross a line sometimes trying to be funny, but you know that your buddy will take care of you, and you'll take care of him at the end of the day. It's like being at home with your family. You may fuss and fight, even physically, with your brothers or your sisters or your mom or dad, but when it comes down to it, they're your family. And nobody is going to come in there and mess with one of my folks.

At the fire department we can be really rough on one another. It's crazy, but we're a family and no one is going to mess with us but us. And we may not be the best buddies with everyone in the department. There are all kinds of different personalities just like anywhere else. And to be honest, I've been doing this for twenty-eight years, and being a firefighter, a paramedic, that's who I am. That's my identity. And as my career is moving along and thoughts of it ending start to creep in, I wonder how I'm going to function when I'm not there at the fire station every three days. I mean, even other people identify me as a firefighter too.

I talk to other people in other professions, and I hear what they tell me about colleagues and bosses, and I hear their complaints about how they get treated. They talk about how unfair it is that so-and-so shirks off some of his or her work. Or this person sucks up to the boss and gets preferential treatment or whatever. And they feel helpless to do anything about that.

I'm not saying that we work in perfect harmony, but we handle things like that internally. We deal with it and address it. And I wonder what it might be like to work someplace else, or spend most of my time away from the people I may be messing around with one moment, and the next seeing how honorable they are.

I may struggle when I leave the fire service because I take people at face value, and I don't play games and go behind anyone's back. I'm part of the big-boy world. If we have a problem with one another, we let the other guy know and do what we can to work it out.

Seems to me, the corporate world isn't like that. And, sadly, some of that is coming into the fire service. I can be outspoken about things, and sometimes that doesn't sit well with others. But I look at it like this: I've been doing this a long time. I've paid my dues. I think I'm a valuable member of the team and I have valuable ideas. You pay me to do this job, and part of the job is to put my two cents in about decisions that need to be made. Those are important decisions, and all I can do is be one hundred percent myself and give my honest perspective. So, if I think something is dumb as hell, I'll tell you just that. That has benefited me in my career and it has cost me in my career. But that's the way I go. I don't play well in the shadows.

★★

Having known Clay for so long, I can't help but laugh at his punctuated emphasis on the fact he will speak his mind. But I think it's important to point out that he doesn't do it to be a disrupter or a contrarian. He does it because he cares, and he has integrity. It may sometimes be easier to sit back and be quiet, but when he can see a problem coming, he feels he has a responsibility to call it out. Even if it's uncomfortable. That's exactly who he is: a lifelong servant who is willing to get uncomfortable, even to his own detriment, to make a bad situation better for those around him. Whether it's being the unsung hero on his best friend's race car, working hard to take a legless buddy duck-hunting, or sitting down with a brand-new firefighter to tell him or her that they really can do this, it just takes some training. He is always looking to help, looking to make the team better, looking to save a life with little concern for his own and all the care in the world for them.

As Clay once told me so eloquently, "You know, I see people at their worst. I see the worst people can do to each other, but I still want to help them. I want to respond and do my job." I think that as long as people like Clay are willing to do this job, "we" will be the lucky ones.

★★

The Choice to Serve

—★★★—

KEITH DEMPSEY

★ Firefighter ★

"I said, 'Hey, Dad. I don't know really how to tell you this, but I want to be a fireman.'"

A DAD'S EXPECTATIONS

I always believed it was expected of me to go into either my dad's or stepmother's business. My dad was what's called an independent publisher's representative. Back in the heyday of magazines, my dad's company sold the physical space on the page to interested advertisers for publishing houses that didn't have their own internal advertising departments. He was basically a contractor for those publishing houses. He and his business partner spent a tremendous amount of time traveling, but they were headquartered in Atlanta.

Obviously, since the advent of the internet, that business doesn't exist to the extent that it once did when I was growing up. They had a great business. At their height they represented about forty-four different publications. My dad was able to retire in 2005 or 2006, so if you plug in the math, he's seventy-five now and retired nineteen years ago as a relatively young man. Obviously, he worked his ass off for it, but this shows the level of success they achieved in business and I'm very proud of him.

My stepmother has been in the insurance and financial services business her entire career. She married my dad after my parents divorced when I was four years old. Early in her career, she was working for a large insurance carrier prior to starting her own insurance and financial services company. She specialized in building large conglomerate policies for athletes, family trusts, and other high-net-worth individuals. For example, when a guy signs a big contract in the NFL or the NBA, they may need $100 million in insurance coverage. Most carriers are not going to want to be on the hook for that amount. As a result, she would put together a policy utilizing a patchwork of carriers until the guy was covered for the desired amount. As a college student, I almost felt it was expected for me to go into one of those two businesses.

My mom is a very talented seamstress. At one point she owned her own fabric store and she still owns her own sewing business. She does custom sewing—dresses and other work. My stepdad was a helicopter pilot in Vietnam and has been in the flooring business his entire adult life. I am immensely proud of all four of the individuals I consider my parents and could not be more grateful for the time and effort each one has invested in me.

★★★★★★★★★★★★★★★★★★★★★★★★★★★★★★★★★★★★

Over the past two decades, Keith has become one of my best friends. I sometimes look to him for an objective view; he's analytical by nature, whereas I can be incredibly indecisive at times and a little too "full steam ahead" at others. But when you get Keith talking about the fire service, the job and duty of protecting your hometown from disaster and tragedy, you see a passion in him he doesn't often reveal.

In my conversations with Keith I saw a striking similarity to the veterans I'm close to. They come in excited and eager, then they start doing the job and see how difficult, stressful, and traumatic it can be. Then they find themselves on the back end of such a career and need an opportunity to decompress and make sense of all they've seen and experienced.

Keith is my wife's brother, my brother-in-law. He also grew up in Dalton, Georgia; he attended the same high school where I met my wife; he had just been there a decade earlier. Like Clay, I invited him along on the veteran hunting trips because I had a hunch that all these people had a lot in common.

Something special started happening on these hunting trips. The combat veterans started talking to and bonding with their first-responder counterparts. There seemed to

> be an unusual but genuine trust that the other might have a very different experience but nonetheless a very similar understanding of life, death, and the stresses of being in the middle of it.
>
> By day's end, I was inspired. I knew there was another story to be told, not one about the warriors defending us on a foreign land, but about the ones saving us here at home every day. Keith was key in this epiphany and now, this book.

★★★★★★★★★★★★★★★★★★★★★★★★★★★★★★★★★★★★

GROWING UP IN DALTON

I was still in single digits when my parents both remarried. I was very fortunate, though. I was the luckiest kid around because I had two sets of parents. Growing up, I was that kid who had parents who got along even though they weren't married anymore . . . definitely the exception, not the norm. In fact, all four of them are close friends to this day.

I grew up just three miles down the road from Joey. He is ten years younger, and we didn't know each other until he started dating my sister when they were in high school. We grew up in the same city, Dalton, but were in different communities separated by a major highway. Oddly enough, my wife, Tracie, grew up nearly across the street from Joey. My sister, Meg, and I often joke we both just crossed the four lane to find our spouses.

Dalton is relatively small, twenty-two square miles in area, and has long been known as the flooring capital of the world. In a previous incarnation, that meant carpeting. Now, as tastes have changed, the manufacturers here have had to adapt to the changing market, and they transitioned to vinyl, tile, hard surface, and laminate flooring. Regardless, there is still a tremendous amount of manufacturing here.

The flooring industry is alive and well and is the primary driver of the local economy.

In the days when Dalton had a population of 30,000, with an additional 100,000 in Whitfield County, it was often said there were more millionaires per capita than in a lot of other places due to local ownership of the carpet mills. I'm not sure where this belief came from, nor can I verify its accuracy, but I heard this numerous times growing up. It created an interesting mix in terms of local socio-economic classes. Today I would say the majority of the local population is blue-collar. I grew up middle class, and I would imagine my upbringing was pretty typical in that regard.

My dad was likely the first in his family to graduate from college. He attended the University of Georgia during the Vietnam War era, and that's where he earned his bachelor's and master's degrees in business administration. My stepdad went to Georgia Tech, and on the flip side of the coin he was in the Army ROTC. When he graduated from Tech, he was commissioned into the Army and served as a Huey helicopter pilot, flying missions in Vietnam in 1970 and 1971.

CHILDHOOD DREAMS

My childhood was at the height of the 1980s G.I. Joe craze, and I was heavily involved in that. We routinely played Army in the woods surrounding our neighborhood and spent endless hours riding go-carts and four-wheelers. I started snow-skiing at the age of four and was fortunate to go on numerous skiing trips across the country over the years. I was also a big rock collector. As a kid, I was fascinated by them. When I was six or seven I got a drum kit. I have no idea what drew me to them, but drumming has always been a part of my life and still is. My daughters would certainly attest to this because I'm constantly drumming on the steering wheel as we're going down the road, not even conscious that I'm doing it. It just happens. I was never formally part of a band outside of school. A few of us in high school got together to play, but it was just messing around on weekends here and there.

Hunting, fishing, shooting, camping—anything outdoors was a big part of my life. I got started in the outdoors with my dad and his friends whose sons I grew up with. We still do a lot of these activities today. My dad and I and several lifelong friends still go on frequent hunting trips together. In fact, we just got back from Argentina, where we were on a wing-shooting trip for dove and pigeon.

I can't express to you in words how fortunate I feel I've always been. I didn't notice any kind of division among the so-called have and have-nots or class divisions while growing up, but I always felt lucky and was conscious that not everyone was as fortunate as I was.

I was part of a pilot program at my high school in which we took a two-year Japanese language program via satellite. This was in 1992, my sophomore year, and for three days a week we were in our classroom watching a TV broadcast with native speakers delivering the lessons from Oklahoma State University. On Tuesdays and Thursdays, we joined conference calls via telephone and spoke with native speakers to develop our conversation skills. I even spent six weeks between my junior and senior years in Japan on an exchange program.

We didn't know what benefit those classes would really have until we got over to Japan and had to survive. We weren't on our own, and in Tokyo and other large cities, a lot of people spoke English, but it certainly taught us what it was like to be truly immersed in a culture, and it helped us improve our skills. Eventually I would minor in Japanese in college.

THE CHOICE TO SERVE

I was also actively involved in my church youth group and played a lot of baseball. Over time, I drifted away from church and the youth group I was involved with. Same with baseball. The sport just stopped being fun. As far as church, I'm not sure why I stopped attending. Perhaps it was my going off to college. I don't have a good reason, and there's no obvious driving force behind it.

When it came time for me to think about college, I had a couple of ideas. I either wanted to major in business at the University of Georgia, or pursue forensic science, and Florida State University had a great program for that. Those options are pretty far apart, not geographically so much as thematically.

Pursuing business made sense, given what I've said about perceived expectations. Forensic pathology was just an interest I developed, and kind of had all along. I was always interested in reading crime thrillers or watching true-crime documentaries. Serial killers and that kind of stuff fascinated me because it was such a departure from the rational world around me. Anything on TV about those topics sucked me right in. Eventually, though, after I really dug into what is involved in being a forensic pathologist, I decided it wasn't for me. Once I got a more thorough understanding of the job, it became clear that I would be doing autopsies or death investigations most of the time and quickly realized I didn't have the interest I thought I had for that career field.

In the end, I decided I should stay home and do some of my core classes at Dalton State College for two years. I transferred to the University of Georgia and majored in economics and finance. Because I stayed home for the first two years of college I was able to volunteer as a fireman, which helped solidify that as a career goal. In the meantime, though, I took a psychology class because my interest in criminal activity was more about *why* they do what they do and *how* they commit these acts. People's behaviors fascinate me.

As you can probably tell from the varied hobbies and interests, it's safe to say Keith had a lot of options in front of him. He was and is incredibly smart and well-read. He has a unique way of being both extremely passionate about something, but also remaining reasoned and well tempered with his actions and expectations. For him, making a decision on a career path at seventeen was a yearslong endeavor of pros and cons. As he admits, he had opportunities because he had successful and engaged parents, and in our little town of mostly blue-collar or outright poor people struggling to make it, that was a rare thing. He was fortunate, but the path ahead and work needed were still his responsibility. I point this out because one of the most interesting things about Keith is that he had the intellect and resources to seamlessly move into his dad or stepmom's business and effectively be a very wealthy man before he hit thirty. But that's not what he chose to do. Instead he honored his parents' hard work by going to a state university and graduating, but then answered a calling he found all on his own. I grew up on the "other side of the railroad tracks" from Keith and his sister, my wife, Meg, though for us it was actually a four-lane highway. But the socioeconomic divide was palpable; not contentious, just obvious. As someone who did not have that kind of financial support or opportunity, I have always admired Keith's career decision after college.

HOW I KNEW I WANTED TO BE A FIREFIGHTER

My interest in the fire service goes back to my stepdad and him being a Vietnam veteran and a helicopter pilot. He has four siblings, and one of them, his youngest brother, is a retired firefighter from my hometown department. Growing up, whenever we had family events, I found myself talking with him, asking him questions about the fire department. I don't recall all the details, but I do know I asked him lots of questions. During Christmas of 1991, we were together at a family function and he told me there was a program at the Dalton Fire Department for teenagers who express an interest in public safety. In those days, these career exploration programs were underwritten by the Boy Scouts of America and were called Explorer Posts. They were designed to expose teens to various careers—firefighting, law enforcement, and EMS to name a few. These programs became terrific recruitment tools for public safety and became life-changing for many of the thirty of us enrolled in the DFD Explorer program.

Former members of that program went on to become not only firefighters, but career military officers as well as law enforcement officers at the federal, state, and local levels. We've got guys who went through the Explorer program who became chief officers in fire service organizations ranging from one-station departments to very large metropolitan ones. That program produced numerous contributors to public safety over the years.

My uncle told me to come down to the fire department and check it out. So, on the day of my sixteenth birthday, I did just that. He told me I'd get to learn about firefighting as a career and get to see exactly what life as a firefighter might look like. I came back from that experience after the first meeting and was "bitten by the bug," so to speak. Apparently, some of the guys in the department saw something in a few of us. I think it was because we were raised right, knew how to act, had a solid work ethic, and we demonstrated an interest in firefighting. For us this wasn't just some place to hang out; we had a legitimate interest in the profession, in turn they poured themselves into us. They instilled in us a genuine respect for the job and taught us how to approach the job the right way.

Now as a thirty-year veteran of the fire service, I try to show up to work every day with an enthusiasm to train the next generation of firefighters because I feel such gratitude, and responsibility, for the time older professionals spent on us back in the early 1990s. By the time I graduated from high school, even with all the other things I was doing, in August 1994, I earned my firefighter certification through the state of Georgia as a member of Whitfield County Fire Department. Whitfield is the county that surrounds the city of Dalton, which is the county seat. As a newly certified firefighter, and because I was an active member of Whitfield County Fire Department, I took advantage of an opportunity to attend the State Fire Academy. So, while living and pursuing a business degree at the University of Georgia, I was also able to complete additional training as a firefighter.

Two weeks after I graduated from UGA, we were at my family's lake house. I'm was on the lower deck with my dad. By late that evening, he and I were a couple of cocktails in and found ourselves sitting in rocking chairs just enjoying the sounds of nature on a warm summer night. I'd had enough whiskey by that point that I thought it was a good time to tell him I wanted to be a fireman. I figure I'm deep enough into the bag that if he gets mad, it's not going to be that big a deal. So I just said, "Hey, Dad. I don't know really how to tell you this, but I want to be a fireman."

He replied, "Hell, I've known that for five years. Go do it and be good at it. Who am I to tell you what's going to make you happy professionally? How could I possibly dictate that to you?"

I said, "So that's it? Wow. Nobody's going to be pissed-off here?"

For years I had built this up in my mind and that was it? In that moment it felt as if a huge stone was lifted off my shoulders. You have to understand I felt an expectation for what I was supposed to pursue professionally. Nobody had ever *told* me what was expected of me, but I *felt* it, only to find this expectation didn't exist.

WHAT WE ACTUALLY DO

It has always been my belief that some members of the public think firefighters are just sitting in the station behind closed doors waiting for a call to go put out a fire. That simply isn't true. By and large, the volume of fire incidents we respond to has decreased significantly over the last several decades. This is obviously a good thing for John Q. Public; it's good for us too, but as *firefighters*, we're not gaining real-world experience and on-the-job training through continued response to these incidents. Our personnel who are assigned to the suppression division work what's called a 24/48 shift, so they're on duty for 24 and off-duty for 48 hours. Therefore, each group of firefighters will staff the station and fire apparatus every third day. Over the course of a year, if they go out on eight to ten working structure fires, they're considered lucky. I know that sounds weird, but you wanted to become a firefighter to fight fires, right? We all want to do what we were trained to do. We're glad that our community isn't burning to the ground. That's obviously a positive thing, but firefighters want to fight fires. As I said earlier, the number of responses where we go out to fight a fire is minuscule compared to the rest of what we're called on to do.

Nationally, medical calls comprise upward of 75 percent of total responses in today's fire service. Automobile accidents are categorized differently from medical calls, though you might provide medical treatment during that call. So it's relatively rare for a fire company these days to go out and actually get to fight a ton of fire. Wildfires are a big concern in other, more arid locations, but in Georgia most fire departments aren't certified for a true wildland response. Instead, the majority of fires we respond to are structural in nature.

Most of us went into this line of work wanting to do one thing, but we are trained for and called on to do so many others. I know the same is true for law enforcement officers. The nature of the job has changed for them too. I think the mental stressors placed on police officers are far greater than those we face in the fire service.

These stressors are compounded even more by the fact that most people love firefighters. Think about it: We can show up at a home that is 100 percent involved in fire and ends up burning to the ground, and people will still thank us for our efforts. These people have just lost everything, and they'll still say, "You boys did a hell of a job, and we appreciate it." On the flip side, cops often bring consequences that people don't like. Many people are made uncomfortable by the punitive nature of what law enforcement represents, so they automatically don't like cops. Honestly, who would want to be a cop in today's society? I admire those who do because I simply can't relate to their motivation to do so, and I think how they are treated and the disrespect shown to them are disgraceful.

Regardless of our career choices, firefighters and law enforcement officers are all exposed to trauma. I wouldn't dare compare what we do in public safety to military service, but when a soldier goes overseas to fight a war, they are exposed to acute trauma over a relatively short amount of time based on the length of their deployment. They're exposed to trauma that is tremendous in quantity and quality. In the fire service, we do not experience that level of exposure in a concentrated time frame. Instead, we face chronic exposures over twenty-five- to thirty-year careers where traumatic events may be less frequent but are more cumulative.

Suicide among public safety employees is climbing, and at an alarming rate. I don't know the exact numbers or how fire service suicide rates compare to veterans or law enforcement officers, but it frightens me that this is occurring within our community. For police officers specifically, it crushes me to think they have the added burden of being disrespected and disliked by so many. I hate that for them. They're a bunch of good dudes, just like we are.

What I've found in talking with firefighters is that a lot of them can't pinpoint the straw that broke the camel's back in terms of mental trauma. They just know that something is wrong. Some guys can identify the exact date, time, and circumstances of a call that really got to them, or specifically mark when the trauma occurred. For a lot of them, though, it's the chronic nature of it, year after year.

★★★★★★★★★★★★★★★★★★★★★★★★★★★★★★★★

This point Keith is making is exactly why I wanted to write this book and tell these stories. Sitting around those campfires on our hunting trips, listening to Keith and other first responders casually tell their stories alongside the war stories we had to tell, I could see the similarities, but also the differences. When you think about Keith's example of a family telling the firefighters "good job" and "thank you" as their house has burned to the ground, you lean toward empathy for the family that has lost everything, but start to think about the firefighter who's being thanked for . . . not saving it? What kind of emotional trauma does that cause him or her? I know well the responsibility of keeping people safe from bombs, but I don't have legs because there was one bomb I couldn't find, and a Marine's wife became a widow because of that same bomb. It's called survivor's guilt. These firemen work tirelessly with the explicit goal of saving property and lives, but sometimes it isn't up to them or how hard they are working; sometimes putting the fire out doesn't mean saving the property on fire or the life trapped inside it.

Now compound such an experience with the fact that most of them are doing it in the towns in which they live, many in the towns they grew up in. When Keith takes his family to dinner, how many of those past fires from his thirty-year career does he pass along the way? How many of them were people he knew or businesses he frequented? How many of them were too far gone to save by the time he and his department got there?

Add on top of that the fact that the vast majority of their calls are medical emergencies. How do you drive by an apartment building every day on your way to work if you've seen people succumb to domestic violence, drug addiction, or a freak accident in that very place?

That's the slow, steady drip of trauma Keith is talking about. The daily experiences of people having a really bad day and looking to you to make it better, coupled with the daily reminders of the times you couldn't get there fast enough or perform enough intervention to save them or their property. That type of trauma, although experienced very differently by different people, is familiar to combat veterans. The lingering effects these experiences have on us are sometimes medically diagnosed as post-traumatic stress disorder, or PTSD.

Unfortunately, most civilians, and even too many first responders, consider PTSD to be associated almost solely with combat veterans. That couldn't be further from the truth. To be honest, after fighting two wars, getting shot at, and blown up, the most traumatic experiences I had were two different instances where children were hurt and I had to respond. Both took place in a war zone, but were not part of direct combat. One child, a little girl, was dead when we got there. This young girl fell from a tree and onto an improvised explosive device (IED). I had to respond before her body parts could be retrieved because planting a secondary IED in the same area was a commonplace tactic of the enemy. The second instance was a very young child that was dipped in boiling water as a means of "getting the demons out" that its parents believed were causing it to cry, when in fact the problem was what a developed world would know as colic. How could you do this to your own child? Those incidents haunt me more than any, and they are directly relatable to what first responders see when they respond to a serious car crash or a domestic violence call.

The fact is, through our bad luck, bad actions, or simply evil intent, first responders see us, the people they serve, when we are at our worst or have been the worst to each other. Then they go home to their own family every night, hoping to forget about it. While in Afghanistan, an *ABC Nightline* reporter asked me what a bad day was. I often go back and listen to my answer as a way of reconciling how I could continue doing that job until I eventually got hurt. When I spoke with Keith, I asked him the same question.

THE BAD DAYS ON THE JOB

I'm going to give you a bad day from the response perspective, then I'm going to give you a bad day from a personal perspective. We go to a dwelling fire, and on arrival we discover a structure that is heavily involved in fire. We execute our initial plan, and we perform a coordinated fire attack during which everybody does their assigned job.

It may not look like it's coordinated, but it is.

A coordinated fire attack may involve several fire companies that are operating independently, with the overall goal of stabilizing the incident. Picture a baseball game where the catcher and the third basemen are doing totally unrelated things when the ball is put in play, but each is working toward the collective goal of getting outs and winning the game.

So, everybody is doing their own thing within this orchestrated chaos, but they have the same short-term and long-term goals in mind. In this particular case, this fire was started by a space heater located in a bedroom where two preschool-aged girls usually slept. Inside this bedroom was a large pile of stuffed animals on the floor. When the fire broke out, apparently the two little girls sought refuge within the pile. In searching the house while fighting the fire, we didn't find those girls. We failed them, and we failed in our mission.

That time period was an extremely busy one, during which we responded to numerous unusual incidents and circumstances. During those few months, we really got put through the wringer. In light of that, the fire chief and command staff believed we deserved to receive a unit citation to recognize our work. The guys on that shift, including me, felt like we were being decorated for failing at our mission . . . for not finding two little four-year-old girls in a bedroom. *That's how they want to handle this? A participation trophy? What's this about? In what universe is that a success? We failed at the very basic nature of our job and they want to hang a unit citation on us?*

Another time we got called out on a motor vehicle accident. It was a Friday night and we arrived on-scene to find two crashed vehicles about 150 yards apart. One was an older-model Dodge Power Wagon, like a Bronco or a K5 Blazer, a large SUV. The other was a mid-1990s Ford Mustang convertible. They were opposing each other in a curve, and when they came together at the center line, the Power Wagon came up over the Mustang convertible and rode up onto what we refer to as an A-pillar. These are the structural members just behind the engine compartment and between which the windshield is mounted.

So the Power Wagon rode up the Mustang's driver's-side A-pillar and struck the driver. With the distance between the vehicles, we essentially had two different incidents working simultaneously. The Power Wagon was occupied by two, midtwenties males who appeared to be under the influence of some type of illicit substance. Whatever they were on, they crossed the center line and collided head-on with a convertible driven by a sixteen-year-old male.

When I got down to the Mustang, the driver, this kid, is in a pair of slacks and a leather jacket, and a dress shirt, and his appearance is absolutely flawless. At first glance, there is nothing wrong with this kid. There's not a drop of blood anywhere, not a hair out of place . . . nothing. He is just absolutely perfect, sitting there in the driver's seat, dead from the blunt-force trauma of the impact with the other vehicle. I don't know why he wasn't cut. I don't know why he showed no visible injuries, but this kid appeared untouched, and this was the night before his prom. That image is burned into my brain. I could pick that kid's face out of a crowd of ten thousand people right now. He didn't get to go to that prom the next night, because he lost his life due to the carelessness of others. That incident took place on a road in my hometown that I don't like to travel on because I think of that kid every time.

Those kinds of incidents stick with you. Again, I consider myself more fortunate than many in having a trusted outlet for these feelings. My wife is a middle school principal and is also forced to deal with difficult situations with her students, so when I have experienced these kinds of things, these stressors, these things that twist my gut, I have someone I can talk to about it. I certainly don't share every detail, but I'm comfortable in sharing and confiding in her far more than many of my colleagues may have an outlet for. There are things she doesn't know about because I don't believe there's a benefit to sharing it for her or me, but I definitely feel comfortable sharing some of these issues related to mental health and the traumas we're subjected to. A lot of the other guys don't have this outlet and end up internalizing all of it. They purposely don't bring it home for fear of burdening others, and that's up to individual preference, but they'll bottle things up and allow internal pressure to build. That's where the mental health crisis begins for many in our chosen profession.

That said, the number one outlet for us to turn to is one another. That's because we know the other guys get it. We've all experienced it, and that in itself is a method of internalization and bottling up, but we do close ranks, and we deal with these issues within our own culture and in our own way. We do talk candidly about these issues inside the fire station. We don't sweep them under the rug. We talk about them, openly, in our own way. That may not be what other people or other groups do, but it's how we handle it. Sometimes our way of working through these traumatic events is simply using a very dark sense of humor. That's prevalent in our business. It serves as a relief valve. We're not very good at going outside our own group because we fear not being understood. I've been party to several after-action reviews and post-incident analyses where outside counselors were brought in, and it's a turnoff unless these individuals have been vetted within our culture. This vetting process helps ensure they actually *do* get it when they say, "I get it."

On some occasions, faith-based or community counselors are brought in to conduct these sessions. They say things like "Oh, I understand what you're going through."

You understand what I'm going through? How many little kids have you seen that have been run over by a car? I seriously don't think you know what we're going through. Don't give me that. Get out of here. That's when we close ranks and shut down to those who would be considered outsiders.

LOSING PEOPLE

Ronnie Dyer and Gary Baggett were two guys at the Dalton Fire Department who helped raise me professionally. They were present in those formative years when I was in the Explorer program. My family knew these guys. They were among those who poured themselves into us because they saw our potential. They knew my family. We were friends away from work. We socialized together. But we lost both of these men in separate motor vehicle accidents when each was in his early fifties and on the verge of retirement. This was a big deal. They were professionals who were well respected within our department and within the wider circle of the fire service. They were both masters of their craft, took the job seriously, and, as a result, approached the job the right way. They chose to invest in us when we were kids, and I can't tell you how thankful I am for that. Both had a large role in molding me into who I am now.

Ronnie Dyer was a lieutenant, a company officer. He was always very active as an advisor to the Explorer program. A few years later he and I became coworkers on the same shift. I got to work for Ronnie for a while as a member of his engine company and was able to benefit from his mentorship and leadership. Gary Baggett was an aggressive and progressive guy. He was an engineer in our department when we first met in 1992 and I was a sixteen-year-old kid. He progressed through the ranks, eventually becoming deputy chief. But he was killed in a motorcycle accident while drawing down his leave time just before retirement. Our A-shift responded to the Dyer incident, but Baggett's crash occurred outside the city, so we may not have had any resources on-scene.

Luckily, I have never responded (or don't remember ever responding) on an incident where I knew the patient/victim/impacted party. I am profoundly grateful, since it is one of those fears that reside in the back of your mind when you live in the same community as you work.

Ronnie was a very interesting guy. He was physically comical, tall, thin, with exaggerated features, a bit like Ichabod Crane. He knew it and played off that. It could be funny just watching the dude move around. He was also quick to laugh at himself. But on March 8, 2006, he was killed when the pickup truck he was driving was hit by a car on the North Bypass in Dalton. Whitfield County firefighters assisted us in extricating him from the wreckage. He was fifty-three years old and had been with the Dalton Fire Department for twenty-five years. The crazy thing about the accident was that he was on the phone with his wife when his truck was hit. She literally heard it happen before the phone went dead. We were all in shock. I was still relatively new in the department, and I remember many of the veterans not being able to talk about it at all.

★★★★★★★★★★★★★★★★★★★★★★★★★★★★★★★★

A few years ago, Keith invited me to an annual duck hunt on Reelfoot Lake, at the border of Tennessee, Kentucky, and Missouri. The shallow lake is quite the anomaly, as it was naturally created by a series of earthquakes in the early 1800s that caused the Mississippi River to back up into a forest. The result is a shallow, swamplike lake featuring giant lily pads and sprawling cypress trees. If you go there anytime in the summer or fall, you'll think you're in the swamps of Louisiana when in fact you're north of Little Rock.

As fate would have it, firefighters from the Dalton and Whitfield county departments started going on this short early-season duck hunt every year on the opening weekend, which usually falls on or around September 11. They started this tradition in the late 1990s, so the double importance of that specific date is more of a coincidence than intention. But for me, a veteran who joined after the terrorist attacks of September 11, 2001, it was surreal to find myself sitting in a boat with two firefighters, Keith and Clay, as the sun came up on this incredibly beautiful and unique swamp on *that* day of *that* month. I couldn't think of a better place or better company to observe it.

I've been fortunate enough to go on that trip several times over the past few years. One day, with a casual comment, as a fireman would make, Keith said, "Hey, I want to show you something in one of these channels that not many people know about." Of course, I was eager and excited, so I jumped in the boat.

We took off midday, between the morning and evening hunts. As we rounded a tight corner in a narrow channel lined with giant lily pads and cypress stumps, Keith killed the engine on his direct-drive boat and we came to a gliding stop against the thick vegetation.

He pointed up into a tree and said, "See that?"

At first I wasn't sure what he meant. I wasn't sure what exactly I was looking for. But then I saw it tucked in the limbs and leaves as if it were an extension of the tree itself. Nailed to the tree a good way up was a T-shirt. "That's Gary Baggett's Georgia Smoke Divers number. This tree is the first spot he hunted up here, so after he died, we marked it as a memorial. His wife gave us some of his ashes and we loaded him into some twelve-gauge shells and shot him out into the lake under this tree."

To some that might seem strange, but to me it made perfect sense. This trip was more than recreation, it was therapy. It was the primal experience of getting away from the "battlefield" in which they worked and lived and having

> a chance to sit in silence with one another and take in the beauty and peace God will reveal . . . when we let Him. For the rest of that trip, I watched in awe as grown men drank a few too many beers, made fun of one another, told stories of glory and heartache, and relieved just a little bit of the pressure they carry inside them. It was a beautiful few days that inspired me to do the same for my brothers in arms.

★★★★★★★★★★★★★★★★★★★★★★★★★★★★★★★★★★

GETTING MEN TO TALK

How do you get people to open up about what has happened to them? That's the million-dollar question. I don't know how you approach it. I don't know how you present yourself to public safety personnel who are by and large a kind of closed society. I guess the more prominent issue is that the fire service is a male-dominated profession, and males are notorious for bottling things up until the bottle cap pops off, then Pandora's box springs open. "Suffer in silence" could easily have been the mantra used by public safety personnel for generations, not unlike the silent approach associated with the Greatest Generation in the aftermath of World War II.

We're starting to get better at opening up within our group. We're also starting to realize that some of the services being offered to us from outside can be valuable. Historically, we've been terrible, but we're starting to get better. Twenty or thirty years ago, I would have given a different answer. When I was a kid in this business, things were very different when it came to talking about anything to do with mental health or exposing personal weakness. I was not educated enough on the culture at that time as I was in high school and not really a part of the fraternity of firefighters, but my impression is that if you were dealing with trauma related to a call you went out on, the response would have been "Suck it up. This is what we signed on for."

Over the last five years or so, I have learned exactly who to go to within our department when looking for someone to speak to a recruit class about mental health issues in our business. There are firefighters I know personally who are strong enough to say, "Hey, you know what? It happened to me, and I don't want it to happen to anyone else. So, I want to share my story." I can't imagine the kind of courage that takes, and one of the guys who has done this repeatedly for the benefit of our members is Clay Headrick.

★★★★★★★★★★★★★★★★★★★★★★★★★★★★★★★★★

"Suffering in silence" is a mantra too many people take up, either culturally or subconsciously, as their own. I want this book to be a conversation starter for the men and women in these jobs as well as a glimpse behind the curtain covering that work for the rest of us who benefit from their service and sacrifice.

★★★★★★★★★★★★★★★★★★★★★★★★★★★★★★★★★

Police of the Woods and Water

JEREMY JUDD
★ Game Warden ★

"You can get into trouble pretty quick being out in the backwoods if you're not careful. And there's no backup that's right around the corner."

A SENSE OF VOCATION

I grew up in Maine. I'm a farm boy, and my grandfather, my family, and just about everybody we knew hunted and fished. I was pretty young, and I was out bird-hunting with my grandfather. He had a camp on First Roach Pond up near Kokadjo, just north of Greenville. We were driving on Frenchtown Road, which runs along the south side of the pond, looking for birds. I was staring out the window. Next thing I know, the truck came to a stop and another one pulled up, then suddenly there was a guy standing at our window.

He asked us what we were doing, and my grandfather told him that we were bird-hunting. I was about twelve at the time. The guy said, "I'll tell you what, I need to show you the difference between a spruce grouse and ruffed grouse." My grandfather and I got out of our truck and followed him to the back of the other guy's truck. He dropped the tailgate and took out two different bird carcasses. He then showed us the difference between these two types of grouse. You could legally harvest a ruffed grouse but not a spruce grouse. He told us that too many times, hunters confused the two species, thinking that they were shooting at a rough grouse when in fact they were shooting and killing a spruce grouse. The number of spruce grouse in Maine is far below that of ruffed grouse, and that's why there was no open season on them. In some parts of Maine, they had become very rare.

When the lesson was over, the man shook our hands, thanked us, and drove off, wishing us good hunting. Back in the truck, I turned to my grandfather and asked who that was. He told me it was Sergeant Pat Dorian, one of our local game wardens. I still remember my words: "Gramp, that's what I want to do. I want to be game warden."

Growing up a country boy myself, hunting and fishing have always been a part of my life. A staple within our southern-friend culture is deer hunting. I don't think I know a man in my hometown who hasn't harvested a deer. But with such an upbringing, poor and uneducated, we didn't have much understanding, or at times respect, for what game wardens do. For the most part, we looked at them as the "fun police." Anytime we saw one was a stressful, scary encounter. Not because we were necessarily trying to break the law, but because we usually didn't really take the time to know what was and was not allowed . . . or legal.

Game wardens have a lot of power and authority. If they catch you in the act of poaching fish, they can take everything you were using to poach with, including your boat. If they catch you shooting deer illegally from your truck, they can take your truck too. So, whether or not we understood their job, we feared them enough to be on our best behavior when they came around.

A good game warden understands this dynamic, and understands that it puts them in danger. In its simplest definition, the warden's job is going into remote areas to search for people who are armed to kill in order to check their license and activities, and if they violate a law, that warden has to cite or arrest them. Most game wardens exercise discernment when appropriate and many attempt to interact with their constituents before they get out in the hunt to remind them that a game warden is in the area or to remind them of rules commonly broken.

Jeremy experienced just that with Pat Dorian. Sergeant Dorian could have stopped them after their hunt,

> checked their harvest, and written them a ticket if they accidentally killed the wrong type of grouse. Instead he gave them instructions and an example to use before their hunt. Jeremy wasn't just intrigued by this interaction—he was inspired by it. So much so that he embarked upon a twenty-year career as a Maine game warden.

★★★★★★★★★★★★★★★★★★★★★★★★★★★★★★★★★★

POLICE OF THE WOODS AND WATER

The fastest way for me to describe what a game warden does is to say that we are the police officers of the woods and water. Here in Maine, we are part of the Department of Inland Fisheries & Wildlife. We often work with other law enforcement agencies, state and local. We don't deal with everything that happens in the woods, but a lot of it. For example, if the police get a call about a domestic disturbance out at a camp, they call on us for backup. Mostly we assist them in getting on-scene. Maine is unique. We have millions and millions of acres that are accessible only by traveling on dirt roads. Not many people live in some of those very rural areas, but a lot of people vacation and recreate on these very remote, difficult-to-access lands.

No offense to local or state police or other officers, but a lot of them aren't fully prepared or equipped to be out in the woods pursuing suspects. They rely on us for our local knowledge of the roads and the terrain. I always laughed a bit thinking about the twenty-one years I've spent doing this job. Maybe two or three times did I ever have to call the police to assist me. Let's say someone we encountered and wanted to process for a violation started to get out of hand: we'll call them in. But I'd say that, on a weekly basis, we get calls from them to back them up. We take over scenes, we take over searches, for example if a criminal flees into the woods. And if you haven't been out in the woods, especially some of the deep woods in Maine, getting in and out of them safely, even without tracking a suspect, can be hard. That's not because they aren't capable of doing it, but because they just haven't been trained. They're in a foreign environment out there.

We get along well with the members of those other agencies. We go through the same criminal justice academy, doing fourteen weeks of specialized training. After that there are another twelve weeks of training on the job, and then you're on probation for a year. So, for those first few months you're in the classroom and doing training and exercises with the trainees who end up in the sheriff's office or on a local police force.

Part of their training is to know their limits. That's common sense, so before any of the other law enforcement people get themselves too deep into the woods, they know well in advance what we do, what our capabilities are, and they contact us immediately. If there are any circumstances that will take getting into the woods to deal with, they call us. And independent of them, we do a lot of work that you might typically associate with police officers. So, along with enforcing hunting and fishing regulations, we are responsible for forms of outdoor recreation. That means people out using snow machines, all-terrain vehicles, boats, and other watercraft. There are regulations related to the use of all those pieces of equipment and we have to enforce them. Unfortunately, sometimes those operators get themselves in trouble and wind up in accidents and incidents. If there's an accident, we get involved in doing re-creations of those circumstances—whether it's a boat crash, an ATV crash, or a hunting incident. We have a group in our offices that specializes in those re-creations. They will come in and map it all out and determine who can get to which place.

I worked for the Inland division, but Maine also has its famous coast, and so we have a Marine Patrol. But as a member of the Inland group, I was out on the open water a lot because our responsibilities included activities and wildlife on all the islands. With all the bodies of water on the mainland and elsewhere, I knew from the start that if I wanted to get hired as a game warden, I needed to figure how to make myself a good candidate. In my mind, that meant becoming a diver.

One of the specialties in our division is a dive team. They are under the large umbrella of search and rescue. We also have a canine division and aviation division, and we do other things like forensic mapping and incident management. So, there's a lot to it. We cooperate and coordinate with other agencies as well.

★★★★★★★★★★★★★★★★★★★★★★★★★★★★★★★★★★★

The shared responsibilities of local law enforcement and game wardens like Jeremy isn't the only thing that forges a good relationship between the two groups. They also share a unique danger in doing their jobs. They don't just respond to the same incidents; they often work side by side in extremely dangerous conditions, and sometimes in dealing with extremely dangerous individuals. The lines of camaraderie were challenged in a way Jeremy could not have foreseen in November 2011. He got a call to respond to an armed man who had fled into the woods. He was tasked to assist in locating him and bringing him in safely.

But there was a catch. It wasn't just any citizen: it was a police officer.

★★★★★★★★★★★★★★★★★★★★★★★★★★★★★★★★★★★

RESCUE AND THREAT

I was on the canine team and we got called to do a search for a police officer who was suspected to be suicidal. I was there with my dog and a few state troopers. One of the troopers, a guy named Adam, also had his dog with him. We decided to take his dog in, even though it was known to bite. My dog had a great record of rescuing people without any incident. I let the other guy do what he thought was best, but I kept wondering if that was the right thing to do. We spent hours in the woods and the dog hadn't found anything. Not a thing. On our way back out and back to our vehicles, though, the dog alerted and darted into the woods. Adam and I and a couple of other team members followed after the dog. Adam and I were in the lead chasing his dog.

We came up on the dog and there on the ground was a tarp. Wrapped up in it was a body. I shouted, "Show me a hand! Show me a hand!" The report was that he was armed. No response. What I could see of his skin looked kind of purplish blue, and I turned to Adam and said, "He's dead. He's deceased." The guy was known to be suicidal and so that was a logical conclusion to make. Then, as I was looking around the scene, out of the corner of my eye I saw part of the tarp move. Then it opened up even more and the guy pointed the gun at us. I fired three rounds and he didn't survive that. I killed him.

It all happened so fast. I mean . . . I've thought back to the time when I was in the academy and when I was baby game warden going through that training. All those different kinds of cops were there, state troopers and others. They gave us game wardens a nickname—possum cops. We were just a bunch of possum cops, right? I think back on the twenty-one years of doing that job, and I wish I *was* just a possum cop, dealing with sick or injured wildlife running around the streets. Unfortunately, I had to do the real thing. The real *things*. That was a really, really tough day. It took me a long time to even be able to tell that story.

You always think about what if you have to kill a bad guy, but you never think about having to kill a good guy. But in this case, was he really a good guy? He was a police officer. But because of that, he knew exactly what he was doing when he raised that weapon and pointed it at us. He knew what the consequences would be. So, where did that leave me? What choice did I have? That left me with a lot to process, and it wasn't until years later, nine years later to be exact, that other circumstances occurred that led me to counseling.

> Jeremy's split-second decision to fire at the man in the tarp was not a decision to take a life. It was a decision to save lives. That's a lot easier to say or show on paper than it is to live with and have vivid memories of. I would imagine Jeremy thought he would probably never have to employ his weapon, much less employ it in such a scenario. But when push came to shove, when lives were on the line, he acted appropriately and heroically. A game warden having to use lethal force is incredibly rare, much less against an assailant who is also a police officer. In fact, Jeremy's incident was the first ever fatality as a result of lethal force used by a Maine game warden. Of course, Jeremy was deemed justified in employing his weapon to save lives by the attorney general's office, but that didn't stop it from haunting Jeremy for years. The trauma of such an incident can take a toll on you.

DIVING FOR THE MISSING

My wife was never a fan of me diving because of the danger. We dive in very deep and very dark holes to look for dead bodies. Dive team members don't bring people back alive. We may bring closure for families of the deceased, but that's it. So switching over to the canine unit was a way for me to find an avenue to provide for my family and be safer so that I could continue to do that. I still wanted to be a major part of the department and give back to the citizens of Maine. Fortunately, that incident in Rumford was the closest call I ever experienced in doing search and rescue. In a way what I was doing for the last ten to twelve years was "safer," but I had to use lethal force that one time.

My parents divorced and my dad lived in Florida, so I learned to dive down there as a kid. I knew that I wanted to be a game warden, and those jobs were hard to get. I knew that early on, so I figured I had to make myself valuable to the team even before I was old enough to apply. That's why I picked up diving. Once I was accepted into the academy in 2003, I was, like everybody else, on probation for a year. Shortly after that, I was on the dive team.

I saw a lot and found and recovered several bodies in the almost eight years I did this. One particular case stands out. This search involved a teenage girl who had jumped into the Saco River in Limington. Maine is huge, at least compared to some other northeastern states, and I had to cover a lot of territory in my district. That kept me away from home. I'd get dragged all over the state too.

This tragedy took place around Thanksgiving. The river was very high then. We knew about where Coreen went into the water but preliminary searches didn't turn up anything. As you can understand, her mother was shattered by this whole sequence of events. One of the bosses, a major who eventually became our colonel, promised the mom that we would bring her daughter home before Christmas. So, a few days before Christmas, I went into the water again. I had been in cold water before, but this was different. It was moving so fast that the cold felt much worse. I remember radioing up to the guys—I have a communications system in my mask—and telling them that I was really cold and that I should probably come up. But I wanted to do my job and find this girl for this mother.

I knew that I was getting hypothermic, and that can be very dangerous. You can get confused and drowsy, and you don't want to be that way while underwater. Fortunately, I got out of the water. I remember sitting on the side of the river, a beachy area, and then I remember some of the team standing and kneeling over me as I was being loaded into an ambulance. I had passed out. My blood pressure had dropped so low that I passed out on the riverbank and they had to rush me to the hospital.

My wife being a dispatcher, they wanted to let her know immediately before anything went out on the radio. They reassured her that I was okay and told her that I had had a little incident. I was fine eventually, and the whole thing was a bit blown out of proportion, but that incident was toward the end of my time on the dive team. I knew I couldn't do that kind of thing to my wife again.

★★★★★★★★★★★★★★★★★★★★★★★★★★★★★★

We know that search-and-rescue crews work hard to find people when a natural disaster or tragedy strikes. It's heroic and perhaps rare to know someone who would risk their life to save someone else. But how often do we consider that there are people in this world, like Jeremy, who will risk their own lives to bring our lifeless bodies home to our mothers? What strikes me about this story is that Jeremy was risking his life to fulfill a promise someone else made to a grieving mother. That exemplifies what a true public servant is. That is the embodiment of selfless service and sacrifice.

★★★★★★★★★★★★★★★★★★★★★★★★★★★★★★

WORKING WITH WILDLIFE

I love the animals and the heritage of Maine and its hunting and fishing. I loved having a job that paid me to protect that. The conservation part of being a game warden is crucial. Without regulation, if we let people harvest animals and fish with no restraints, there'd eventually be no hunting. The populations would just be decimated. So keeping people honest and maintaining that heritage, that privilege we have in this beautiful state has always been important to me. I love being out in the woods, and being paid to do that? What could be better?

Game wardens out here within the hunting community are treated with great respect. When we met that warden when I was a kid, my grandfather, Larry Judd, showed respect to the warden and the warden showed respect to us. There was nothing confrontational at all. My grandfather liked the idea of one of his grandkids being a game warden. Once I showed an interest in the career, he did a lot to support me. Besides learning to dive, I was able to take all kinds of classes—hunter safety, ATV safety, guiding—all with an eye toward improving my chances of getting hired. To give you some idea of how coveted those jobs were, when I applied there were nine hundred others who did too. Only seven of us made it.

I owe a lot of that to my grandfather Judd. As for my dad, even though he and my mom were divorced and he was down in Florida, he supported me too, just like my mom. My mom, Pauline, remarried, and she ran her parents' farm for quite a while. We all helped out—my brother Mark and my sister Christine—and we ran around and had a great childhood on that hundred acres. When her parents came back, they took over the farm again. They had a huge herd of milk cows, and my mom was especially into horses.

★★★★★★★★★★★★★★★★★★★★★★★★★★★★★★★★★★★★★★

Earning the job and title of game warden isn't simply a matter of applying and being one of the lucky ones chosen. Just as Jeremy learned to dive and how to hunt and operate an ATV safely, he also worked on his formal education. Having come from a very humble socioeconomic background, he had to work hard to stay enrolled in higher education, but he eventually earned an associate's degree in criminal justice. He also attended the formal police academy before being specifically trained for the work of a game warden.

THE THINGS YOU LEARN AT GAME WARDEN ACADEMY

It's been a long time since I was there, but I do remember a lot of the scenarios that we ran through. They called it mock training, and we did practical exercises on things like the proper way to execute the mechanics of restraint and arrests. It was all hands-on stuff, like how to place someone in handcuffs. And we learned how to fight, including self-defense techniques. You hope you never have to use any of that, but some folks don't want to be taken in. Basically we had to learn how to subdue people the right way. We fight by rules, while criminals never do. We even learned to run traffic and radar. I never did that in my career, but regardless it was good to have a foundation in all aspects of police work. We often were called in on backup by the police, and it was helpful to know what they were doing. Over the course of my career, I'd say I arrested about one hundred people total.

But where the road diverged from standard police training was in those next training days at the advanced warden school. A lot of that was built around conservation law enforcement, but more important was the search-and-rescue component. What underlay that was another basic principle: how to be independent and work on your own. Some of it was similar to survival training. You can get into trouble pretty quick being out in the backwoods if you're not careful. And there's no backup that's right around the corner, so that kind of training and those exercises and experiences are essential.

We did a lot of our search-and-rescue exercises at night. It's one thing to be out there in daylight, it's another thing altogether in the darkness. Some of what you learn is tactical but some of it is just plain old physiology. Your body needs rest, and around four in the morning is when it really tells you to shut things down so that can happen.

You have to learn to overcome that. And so, many times we'd be out there much of the night and finally roll back into the academy an hour or two before sunrise. Then it was right into physical training at 5:30. You had better get your quick nap in, because before you knew it, you'd be out there again running six miles. But the training was useful. I can't tell you the number of times when I was out on a night mission, then got back home just as the sun was coming up, and got another call and had to head out again all sleep-deprived. So being pushed so hard in training was good. It showed you what it was going to be like once you were out there in the real world.

DEALING WITH DEATH IN THE FIELD

Some work, though, you really can't get trained for. That includes dealing with dead bodies. I remember talking with friends, and they always seemed to be interested in stories about recovering people. For me, it had become a kind of "Yeah, whatever, it's just another day" kind of thing. My daughter ended up becoming friendly with a girl in her daycare. My wife and I got to know that young lady's parents. They were a few years younger than us, and during one conversation, the subject of dead people came up. Neither of them had ever seen a dead person. They both had both of their grandparents and parents still alive. They had never been around death. I couldn't imagine that. And they couldn't imagine what it was like for me to have seen dozens and dozens of dead bodies in my life. That was what my job involved. But in training, no one ever talked about that being a part of our reality when out in the field and working as a game warden. In time, you get so accustomed to it that you forget that other people aren't.

One time down in Saco a couple of agencies had been looking quite some time for a young man who had gone missing. Eventually I got the call to head down there with my dog. They had been using their own dogs but without any luck so far.

A fairly new game warden went down there with me. We started out in a mixed area of fields and woods. I let my dog out, and she threw her head back and took off. I thought this was a really good sign. We went after her and she led us right to the body. He wasn't that far from the houses in the area. He had gone out in back of them and hung himself from a tree.

One thing I remember even more than the sight of that body hanging from a tree was the look on our new guy's face. He couldn't have been with us for more than a month. He was just standing there, ashen and unmoving.

I told him that he didn't have to do anything but stay there. I told him that the body wasn't going to move and that he, the new guy, was going to be fine. This wasn't going to be a one-off for him. This was a part of the job. Maybe nobody else had communicated that to him before, but that was the truth of it.

I didn't say this to him, but I wished that somebody had told me that it was going to be a part of the job. Maybe because of the specialties I pursued—diving, search and rescue—my experiences weren't that typical. Still, it would have been good to know that there was going to be some gruesome things I'd have to see. Early on, I was part of the team investigating a boating accident, and I was doing a diving recovery of two bodies. A big boat completely ran over a smaller one. The man and woman in the small boat had gotten killed. The water was a bit murky, and I saw something. This was a couple of days after the incident. What I saw was her severed head, and her tongue sticking out of her throat, straight up into the air. The guy had been hit in the back and split open and then gutted by the propeller.

It took me a while to develop the ability to treat looking for a dead body like I was looking for an object. That's no disrespect to the person, but that was the way I could handle it without getting too emotionally wrapped up in it. I'd tell myself that I was out there looking for a tire that fell underwater. Otherwise it would be too hard to do that job. It's sad. It's too bad. I hate when people have lost a loved one. But my job on the dive team was to recover that body. That was how I could help the survivors. I had to learn how to do that, and how to help myself and others on the team to do that job without getting all torn up about it.

I remember the first body I brought up. The boy was fourteen years old and had been showing off for some girls when he drowned. The water was too cold and he locked up. I can picture myself at that age, or older, doing the same thing. It was a very tough situation, not because it was my first, but because the boy's family was right there watching us. I was there with Mike Joy, the head of the dive team at the time, who sent me into the water. I slid down under the water, and the first thing I saw was a big log. The end of it seemed to be glowing. I made my way over there, and there was the boy. I hadn't been in the water but three minutes when I found him.

I had him in my arms, but I didn't want to bring him up to the surface with the whole family watching. They didn't need to see that. I didn't need to see them seeing that. Mike had me bring him downstream a ways from the family. I did that and brought the young man out on the opposite shoreline from where the family was. The medical examiner came in to look at the body and formalize the cause of death. Eventually the family was allowed to see the young man. There's a whole process and procedure to all this.

I quickly realized that for me, part of the process was to get away as soon as possible from the scene. I didn't want to be around the family. Other dive team personnel feel the same way. I didn't want to see them sad. I didn't want to feel that sadness. We just want to get our gear together, get back with the other members of our team. Because they understand what we all just went through. They get it that we were just doing our job. It was just another day, just doing what we do.

Over time, that takes a toll on you. Still, it's a huge honor to qualify to be on the dive team doing that work. We get a lot of praise. The community appreciates us. Even within the agencies, we get a lot of respect. A lot of people want to be on that dive team. It's a cool thing to do, and the team is very tight-knit. That bond can get you through a lot of the dark stuff. In the end, it was a hard decision for me to leave the dive team and move on to the canine team.

★★★★★★★★★★★★★★★★★★★★★★★★★★★★★★★★★★

Jeremy told me very matter-of-factly, "When you're diving, you're not rescuing; you are recovering. You only find dead people underwater. But when you have a dog, and you're tracking a scent, there's a chance you can save a life, and that was important to me. To find someone alive." That was a new concept for me.

One term used in military and civilian emergency response is "rescue and recover." I had heard it a lot. I trained for it and have even been a part of rescue and recovery missions in combat, but not until hearing Jeremy explain it did I conceptualize that with that one word *and*, we go from life to death. We go from hope to sorrow, from a hero's welcome to the look of pure devastation on a family's faces. Jeremy didn't just learn this difference in language; he experienced it time and time again. Knowing

> the toll that diving was taking on his body and mental health, I don't think it's a stretch to say that moving to the canine unit was about saving a life, perhaps starting with his own.

A "ONCE IN A CAREER" DOG

I got a black Labrador retriever pup from another game warden. I got her because I wanted her to be a hunting dog, mainly to retrieve ducks. Her name was Tundra, and the first season I had her, I trained her. I read all the books and followed the experts' guidelines. From the start, I saw that there was something special about this dog. It was weird how fast she picked things up, because all I had to do was show Tunny something once and she got it. If I showed her twice, she would actually get mad at me. It was like she was saying, "I got it. I got it. Let's move on to something else already." In her first year, she performed 128 duck retrieves. That number is just unheard-of. I remember thinking that this dog could do something even more valuable, more important.

In 2009, I took her to our canine team and asked them if they could evaluate her as possible candidate to do search and rescue. Typically, they won't do that if one of the game wardens owns a dog. The way it normally works is that a team member gets assigned a dog that has been brought in and trained. You don't work with your own dog. I brought her to an obedience training certification class. Some people refer to this as the dog being able to demonstrate their manners—basic obedience—sit, stay, heal, come, recall at a distance after being away for five minutes.

I watched the other dogs. Some handled it with no problems, while others failed miserably. This was the initial screening process for the dogs and their handlers, so there was a good bit of tension in the air. Roger Gay was the head of the canine team, and after observing this certification exercise, I walked up to him. I bet him that my dog could pass this test today. Keep in mind that she was barely over a year old and it often takes years for a dog to learn all these behaviors.

He told me that if I thought so, I should go grab her. I suspect he was thinking that I was just bragging on her and that she'd fail. Turns out she passed with flying colors. All dogs are motivated by something, and you reward them with that—food or a ball, mostly. Tunny was a food dog, but what really motivated her was praise. She really wanted to please me. I was really happy that she earned that certification. I took her to another phase of training. This took place over three days, and by the end of it, she had done one of the main elements, what we call puppy tracks, almost to the level that she could have been certified then and there as a tracking dog. One of the other handlers told me that if you're lucky, once in your career you'll come across a dog like Tunny. He encouraged me to apply for Tunny to become a tracking and evidence search dog. I asked if she would be accepted and was told yes, no doubt about it.

So I sold Tunny to the state of Maine for a dollar. Regulations required that she be owned by the state. I'd keep her with me and the family, but the state would pay for her veterinary care and food and other expenses. That's how Tunny and I became canine team members. My dive team boss wasn't too happy about me leaving. To be honest, I wasn't either. Not fully regretting the move, but still wondering if I'd done the right thing, I was all over the map with my decision.

But something would soon happen during that period of doubt that would help confirm for me that I'd made a good one.

I got a call from one of the canine trainers to come up to Belfast, Maine. A search was on for an Alzheimer's patient who had gone missing. I arrived on-scene and learned that the search-and-rescue operation had been going on all night. At the rally point, I looked over the map. We were at the house where the man lived, and they'd already searched a rectangular area just behind it. This was our first real call, and I figured that Tunny had some jitters in her.

I didn't put her GPS tracking collar on her, but I turned her loose so that she could get some exercise and get those jitters out. I let her out in the backyard, which sloped away from the house and then down into the woods. I was there with Tunny when all of a sudden she threw her head back and took off.

I chased after her, but pretty soon she got small in my vision, she was so far ahead of me. As I said, she was without her GPS tracker, but I didn't want to call her back because it was pretty clear that she was on to something good. I couldn't see her well, but I could hear her barking and followed those sounds. We were heading toward Route 3 in Belfast. The construction company had put up a wildlife fence that ran along it. When they did that, they pushed a very large rock down a small slope to clear the way for the fence. But the fence was touching the rock and at that intersecting point there was a big hole opened up by how the rock had moved. The man must have been walking along, hit the fence, then slipped down into the hole, becoming trapped. He was stuck in there and that's why no one had found him.

That's where I found Tunny. At first I could just hear a muffled barking, but when I climbed in the hole, I saw that she was standing on the man's chest. I pulled her off him. I had to grab the man by his boots to get him out of the hole. He was disoriented and shivering from the cold. But he was safe and alive.

I was practically shivering with excitement. I was losing my mind with happiness. Tunny and I were both rookies, and I had let her out just so that she could run off whatever anxiety she was picking up on, and she found this man within a few minutes! I made the call in and Terry Hughes came tearing down the road. I flagged him down, and a little while later an ambulance showed up and the man was loaded into it. All of this took about a half hour; I stayed on the scene. Adrenaline was pumping through me. I was playing with Tunny, rewarding her with praise, and feeling great. This wasn't like it was with the dive team. We had rescued a living person and not found a dead body.

Once the man's son arrived on-scene, I walked over to the ambulance. The father was sitting upright in the ambulance, and it seemed clear that he was in good shape. He was talking with his son. Turns out he was a prominent man in the area who owned an oil company. After a minute, the son turned away from his dad and came up to me. Tears were streaming down his face. He extended his hand to me. He couldn't speak at all. We just shook hands, and I knew he was thanking me. He patted Tunny a couple of times too. That's when I knew that I had made the right decision in leaving the dive team. It just seemed meant to be. We had found that man in the area that had already been searched. Who would have thought he'd be under a big rock, invisible? But Tunny picked up the scent and found him. She could do what we couldn't. That man would have died if it hadn't been for Tunny.

★★★★★★★★★★★★★★★★★★★★★★★★★★★★★★★★

For the next decade Tunny and Jeremy worked together and all told, they found twenty-two individuals. K9 Tundra, as she was officially known, and Jeremy earned two K9 Search and Rescue Case of the Year awards, in 2011 and 2015. Along with those rescues, the two of them were responsible for locating critical evidence in many cases that led to successful prosecutions of offenders.

In 2019, K9 Tundra won the first North American Wildlife Officers Association Loyal Partner Award for their lifesaving search and rescue and other contributions in fish and wildlife conservation law enforcement. As Jeremy understood before any awards, Tunny was truly a special dog. But in 2011 everyone found out just how special she was when she won the Maine Criminal Justice Academy's Iron Dog Award, a multiagency competition for the canine team that has proved to be the most successful in evidence detection. Jeremy does get some of the credit. It took courage to leave the dive team and it took a lot of love and work to prepare Tunny to use her gifts in such a rewarding way.

Although every rescue is rewarding, Jeremy and Tunny especially enjoyed helping find and rescue some of the most vulnerable among us. One of those cases was featured on the Animal Planet show *Northwoods Law*. From 2012 to 2017, the reality TV show featured a number of Maine's game wardens at work.

★★★★★★★★★★★★★★★★★★★★★★★★★★★★★★★★★★★★★★★

TUNNY'S REPUTATION GROWS

At this point, the production company was done filming for that season. They had hoped to be able to get footage of a search-and-rescue operation, but that hadn't happened. I had become friendly with one of the field producers who was often out with me. We were at my camp before he was going to head back into town and the lodgings they had for him there. I got a call about a search-and-rescue operation that was in progress to find a man with dementia. This was in November, when it was getting very cold and the daylight was short. The producer, Jameson, asked if he could go along with me. It turns out it was a good thing that he did. Not so much for the show, but for the sake of this lost guy.

The search had been on for a day by the time Tunny and I arrived on-scene. I walked Tunny down this path. There was a good bit of moisture on the ground and the path was a mess of footprints from all those searching the area. It would be nearly impossible to track the missing man's prints. That meant we'd be relying on Tunny and her senses. I kept thinking, *C'mon, little dog, we really need you on this one.* The path followed a brook that was heavily choked on either side by brush. Tunny was out front a bit, and then she bolted over the bank and into the brook. That wasn't typical for her, so I followed her through this tangle of alders, working downstream. Eventually I found Tunny and the man. He had fallen into the brook and was caught up in the alders and couldn't extract himself. His head was above water but he was still in it partially, and it was November so it was freezing. He was a big guy, and I couldn't get him out on my own. There was also no cell service there, so I sent Jameson back for help.

In the meantime, I tried to keep him calm, doing as we were taught. "Hey, Ben. You're going to be all right. Talk to me, Ben. Talk to me."

He didn't say anything, just looked at me. By this time, one of the other wardens out there had joined me. He let me know that the guy was nonverbal. He couldn't hear and couldn't speak. I hadn't realized that. It was bad enough to be out there lost, but oh my God, to not be able to hear or to shout out for help? Of course, later the guys gave me crap about my talking to him when he couldn't hear me or respond, and it was funny to us how the whole situation went. But that was just how we dealt with all the bad things we had to see and do. We got the man out of there eventually, and the TV people were happy because they got the story they wanted. But the only reason they got it was the permission of the family, and the family would only let them have permission if the production company met their one demand. They wanted to meet Tunny and me.

We ended up going back to that area for the meet-and-greet. We met the man and his wife, and they were very grateful. When I was out there on their farm, a few guys were in a barn pulling out hay to feed the livestock. At one point one of them came over to me. He introduced himself and said, "Isn't it funny that we're both here?" I didn't have an answer for that, but then he filled in the blanks. He asked if I remembered Bucky Coots. I told him that I did. Turns out I was talking to the son of a man I had rescued from drowning a few years earlier. He was out ice fishing, fell through, and survived by hanging on to a piece of ice floe as he drifted across the lake. We got an airboat and went out and rescued him. That was humbling, to be there with two people whose lives you played a big role in, and then to have someone else there whom we'd also helped. That was cool.

Another time, we got called out again to come in after another dog had failed. By that time Tunny had developed a reputation as a kind of go-to canine. If other attempts weren't working, bring in Tunny. Sergeant David Hall was with the Cumberland County Sheriff's Office for thirty years. His team had been called out but hadn't found the young man. The boy had some learning deficits and other cognitive delays, and he was in a residential facility but had been home with his family for the weekend when he took off.

I remember Dave walking up to me and seeing Tunny and saying, "Oh great, *another*." He'd clearly run out of patience with dogs, even though he was joking a bit because we know each other pretty well.

I like a challenge. I looked at him and said, "Listen, I'll be back in ten minutes with your boy." After I interviewed the family, David said, "All right, Judd, your clock is starting."

Tunny and I zigzagged through the yard for a bit before she took off. Another warden and I began pursuit of him. The boy's parents had planted the seed in my mind that their son might get violent if we caught up to him. We would likely only be able to get him to come along back with us if we handcuffed him. He was a big kid—six feet, six inches tall, burly—and did not want to go back to that facility.

Well, sure enough, it didn't take us long. The young man must have heard Tunny and been flushed. He took off running. In my time apprehending people, I had seen the bad things adrenaline can do to someone who doesn't want to be caught. I'd also seen the good things it can do. But by taking off running, he was leaving a scent trail of adrenaline, making it easier for Tunny to track him.

She caught up to him fast. We came up on the two of them in a brush pile. Tunny was on top of him, and I was prepared for a fight. My adrenaline level was up too. I was yelling at him to show me his hands. I had no idea if he was armed, and you never take any chances with that. So I piled on top of both of them in that brush pile, me breathing hard, him breathing hard, and Tunny breathing hard and barking. From the outside, it must have looked like a big pile of commotion. Dave, the other warden, caught up to us, and he was the calming presence. He got me to get off the boy, and rather than being the belligerent bear to deal with, he was a teddy bear. The boy started crying and hugging Dave and asking him to not let them take him back to that residential facility. He wanted to be at home.

Of course I felt horrible, because I'd been led to believe he'd fight us and he didn't react that way at all. But we sat with him and walked him out. Dave acknowledged that I'd won the bet. "Seven minutes," he said, because he'd kept an eye on his watch. "Are you effing kidding me?"

Dave appreciated what Tunny did. His own people had canines, but anytime they needed a search-and-rescue dog, he would do what he could to bypass them and tell his people to call me and get Tunny on the job. That was really gratifying.

I'm grateful that God chose to give her to me and to give me the patience to train her. I'm not a patient man and she taught me a lot about patience. I have to go, go, go all the time. But she was slow and thorough. I just had to put her on the ground and wait and let her do her thing. I wasn't in full control and didn't have to be. That was a lesson I needed to learn. She was going to make me look good in the end if I trusted her. Tunny earned this positive reputation in our public safety community, but also in the wider community.

I love the Maine Warden Service, and I'm proud of the work I did. I was recognized for the good service I provided to the people whose tax dollars paid me. I even was asked to go to Texas to help provide flood relief. Nineteen different times in my career, I received awards and citations, appearing before governors in full-dress uniform, the whole deal.

A GOOD DOG

I remember thinking of Tunny during her career and wondering, what am I going to do when something happens to this little dog? She was a special dog to us, but also to the community. There were always stories in the newspapers about the awards she won, and about the rescues. Like I said, I took her to schools with me to do talks. I remember getting up one day. She was lying at home. I thought that something didn't look right with her. She acted kind of like a drunk person, unsteady and stumbling. I thought, *This ain't good, girl.* I took her outside and she seemed fine. I told myself that this wasn't an emergency and that I'd take her to the veterinarian's office the next morning. That morning I got up and it was like she was a pup again, so I thought, *Where'd you come from?* She was excited and happy, and we spent the whole day together. I was having a rough time then, and she was there for me, making me feel good. But the next morning after that, I could see it in her eyes. Something was really wrong. She couldn't stand up.

I took Tunny to the vet, one who worked with our agency and was a good friend. It turned out she had a tumor in her chest and it had burst. The vet told me that we could do an emergency surgery on her. No guarantees, but maybe she'd live another six months. I asked him what he would do. I reminded him that he knew how special she was, what she meant to my family and everybody else. He said the operation would just be delaying the inevitable. He believed it was time to say goodbye. I agreed. I called my wife. I still get torn up even talking or thinking about this. But she came down to the vet's office with the kids. We sat with her and we put her to sleep. It was hard, but it was the right thing to do.

★★★★★★★★★★★★★★★★★★★★★★★★★★★★★★★★★★★★

Clichés like "Only the good die young" or "We just don't get enough time" are never more accurate than when spoken in memory of a good dog. This interview with Jeremy came at a serendipitous time for me. About a week after this conversation, I was at a theme park in Florida with my family enjoying the end of summer when we received a call from my in-laws, who were keeping my dog Tucker. He had lost his appetite the week before and the vet sent him home with a very ambiguous prognosis. All we knew was that he was having problems with his liver. However, the call brought terrible news. Tucker had taken a turn for the worse. The veterinarian got on the phone and proceeded to tell me our good boy was out of time and that he wouldn't live another twenty-four hours. It would take at least twelve hours to gather the kids, pack up, and drive straight to him. She said he was in a lot of pain, so I gave her the green light to help him pass with ease. I'll never forgot that day. I was sad he was gone and felt a tremendous amount of guilt that I wasn't there for him.

But that's life. Like so many of the stories Jeremy and others in this book tell, life isn't batting .1000 in respect to all our decisions. It isn't about never making mistakes or always being in a situation where we can do something to make it better. Sometimes the circumstances are dictated for us, and all we can do is respond and learn.

Tucker passed peacefully. We finished our trip and now he rests in a pine box next to a picture of us when he was a puppy. Life, quite literally, goes on. For Jeremy, Tunny was more than a dog or a friend; she was a way for Jeremy to move forward in his career. A way to go from reclaiming bodies to rescuing lives. We all need that reminder in life: when things seem dark and dismal, there is still redemption, forgiveness, or simply happier days ahead.

THE PRICE WE CARRY

It came with a price. It got to a point when the stress of the job and some things I can't talk about got to me. I was sleeping just one or two hours a night. I ended up going to see my doctor and he said that I was suffering from PTSD. He gave me some medication to help me sleep. I also started to talk with a couple of counselors. I could talk some with my wife. She understood as best she could, but it just wasn't in me. It didn't feel right to burden her. The counselor had some idea of what we did, but when I related some of the stories I've told here and a few more, she said, "No wonder you're exhibiting these symptoms." For example, the year before the cracks started to show, I was an acting sergeant. I was managing a division and sometimes had six or eight people under me.

We were dealing with five fatal crashes involving snowmobiles. We were working one of them, and it was sad as well as out of the ordinary. A woman driving a snowmobile was out with a group when she hit a pressure ridge—where the snow has been drifted up by the wind and other environmental factors—crashed, and died. Everyone else with her fled the scene. So we were working that case. Then I got a phone call from the state police barracks. They had another one. A nineteen-year-old boy was drunk out on one of the islands. He had decided to go for a ride on a buddy's sled, hit a pressure ridge, and died. In another one, the first officer on the scene was Sarah, a young woman who had just started working with us. When I arrived there she was pale white, like a sheet. I asked her what was going on and she told me that the victim was female and eight and a half months pregnant. She was dead. The baby was alive. We were way out in the woods. The baby was going to die if someone didn't do something, and the woman's husband wanted us to be the ones to do something—cut the baby out. Sarah related this to me, and I told her that we couldn't. We were not trained to do that. We did rush the woman to the hospital, where doctors performed the surgery, but the baby did not survive. Sarah was upset because she wanted to have kids someday and had to witness this scene.

The next morning, we got a call about an ATV accident. A young man was released from prison and went home to his mother's house. There was some legal reason, such as an injunction or protective order, that prohibited this recently released prisoner from having any kind of communication with his brother, who also resided at that address. The mother, knowing all of this, forced the "bad son" out of the house. He had to leave. He did, but he also left behind a suicide note. He told his mom he was sorry. He told her that he loved her. He told her that he was going to kill himself and asked her to let the police know to come and find his body. She woke up the next morning and found this note. She got her other son out of bed. He was angry and hopped on his ATV to start searching for his brother. He lost control of the vehicle, hit a tree, and died. So now the mother had one son who had likely killed himself and another who was dead on the trail. She was the one who had discovered the wreck and her son's body.

When I got there, she was on the ground feeding herself dirt. We had to get her out of there, process the accident report, and get the search going for son number one, who left the suicide note. But in the end it was an empty threat. We found him passed-out drunk in his truck in front of a buddy's house.

★★

> Jeremy is a proud veteran of the Maine Warden Service and claims one of the most decorated and storied careers in its long history. His time with the Warden Service is coming to an end, though. He is bittersweet about this, but as they say, all good things must come to an end. For Jeremy, though, it's not the uniform or the badge; it's waking up every day knowing that what he does matters. Knowing that his time spent learning and honing skills was for a higher purpose. I met Jeremy while in Maine on a once-in-a-lifetime moose hunt. I was invited by Dave Hentosh of Veterans Afield Foundation, and Jeremy had volunteered to help guide wounded veterans on moose hunts on his days off. By the end of that hunt, Jeremy had carried me piggyback across a shallow river and up an embankment. He had carried me on a stretcher across a swamp. He had held my legs out of the water while I tended to a sore on one of my nubs. He had given of his own time and effort to help me do something I'll remember for the rest of my life: harvest a Maine bull moose. I know that in or out of uniform he will continue to help people in need and use his skills to make the world a better place.

★★

PART 2

ENDURING THE TRAUMA

Everything that we do, we do as a team. So, if my guys on my truck or my team, if one of them has a problem, that's everybody's problem, because we either have to take care of them, save them, rescue them, or, at the bare minimum, we have to pick up their workload.

—*Clay Headrick*

The camaraderie with the guys on the team helps us all so much. The relentless ball-busting that goes on is our love language. These guys are good at a lot of things; they're all high achievers . . . and they're also kindergartners. They have the ability to perform at a high level and then descend into childlike antics. They can flip that switch.

—*Tommy Wehrle*

When we finally got to them, the thing I remember most is the big hug I got. A few of the guys were nearly crying, saying, "Thank God you're here." I knew that my job was something that I could be proud of. Being able to help people in need, guys on those teams are very special, because we're willing to risk everything to save a life.

—*Vincent Vargas*

Family of Service

KATELYN KOTFILA

★ Deputy ★

"That's the way that I honor them: by making the best of my life and doing the things that they can't do. So, if they're looking down, they'd be proud."

A FAMILY TRADITION

In some families, the pull toward an occupation is pretty strong. In my case, the tradition of family members working in law enforcement is a huge deal. Most immediately, my father, John, retired from the Massachusetts State Police as a sergeant. His father was a local police officer in western Massachusetts, and his wife, Terry, was raised in a cop house. That tradition wasn't limited by geography. Mom's brother, Uncle Bobby, served in Pinellas County, Florida, in the canine unit, and my cousin now works in the sheriff's office in that same county, which includes Tampa and St. Petersburg.

For me, growing up in a cop family was a simple equation. Dad worked as a homicide detective. I thought it would be really cool to see a dead body and that it would be even cooler to do the job of solving those horrible crimes. Part of my fascination was fueled by the work Dad did; a large portion of it was fueled by watching television shows with him. Watching *Criminal Minds* together was like watching a football game with a former pro quarterback. That's the kind of insight and analysis he could bring to the show. I remember Dad let us know what was real about the investigative work presented in the show and what wasn't.

Life at home in Falmouth, Massachusetts, on Cape Cod, wasn't all murder and police procedurals. We also watched *South Park* and reveled in its irreverence, much to Mom's dismay. But we bonded over that experience, and I loved how honest and funny the show was. The characters were able to say the things that other people were afraid to say.

Those experiences watching those two very different shows reveal the complexity of what it was like living in a household with a law enforcement officer. As much fun as *South Park* was, it was the same mix of reality and TV unreality as *Criminal Minds*. As much as Dad liked being the one to indulge the kids by watching a show with foul-mouthed animated kids providing lots of laughs, he was also the family disciplinarian. I spent so much time being fearful of failing to live up to the standards that Dad set for us and the resulting punishments when we failed to clear that bar.

I remember him saying, "You do know what I do for a living, right? So, you can't lie to me. I wouldn't bother trying."

I thought to myself, *Yeah, he's right. He interrogates people for a living.*

He used to tell us that he would take a piece of hair and put it across the front door and its frame. That way, if it got displaced, he would know that we had snuck out. I couldn't resist the temptation to see what would happen, and early in my high school days, I did sneak out to meet a boy, and sure enough, five minutes after we had met up, my phone was ringing.

Dad told me to get my ass home right away. I did, and Dad was not a happy camper.

It's funny now. It wasn't back then.

So, yes, I felt the pressure of being the child of a law enforcement officer. Dad let each of the kids know that it was different for us than it was for our peers. It wasn't like we were a dentist's kid who got in trouble and made news on the neighborhood gossip lines. We were cops' kids, and cops were all about enforcing laws and right and wrong. How would our behavior reflect on him if he couldn't maintain law and order in his own home? What would people think?

He also made sure his kids understood that there was a time and place for crying. If you were injured, then tears were okay. Mental or emotional pain? No. Those were alligator tears and not to be let out. I learned how to control my emotions and how to shut them off completely when necessary. That would serve me well later in life, until it didn't serve me well at all.

I don't think I ever heard my dad talk about the emotional aspects of his work. I think that's probably related to his upbringing too. My family is really good at avoiding especially emotional stuff. We just avoid it, pretend it's not there, and just kind of shove it back down. I probably learned from my family that if you have an emotional issue, you just pretend like you don't have it, and everything's fine. We're good at that, and we're good at laughing about things. Having a dark sense of humor helps. I learned that from my dad, and I've gotten pretty good at using it. I used to joke with my parents that I wanted to study psychology so that I could figure out what was wrong with them.

★★★★★★★★★★★★★★★★★★★★★★★★★★★★★★★★★

Hearing Kate describe her upbringing sounds almost too familiar. If I didn't know her, I'd think she had to be a combat-experienced Marine, with her penchant for dark humor and impressive emotional intelligence. Perhaps the similarities between combat veterans and law enforcement officers go beyond weapons handling, uniforms, and lingo. I know many service members who wore the uniform because they got the inside track; they grew up with a mom or dad or both who were serving. They grew up with a standard that other kids didn't have to abide by, or at least not so strictly.

But even without the pressures of having a parent in uniform service or law enforcement, so much of your drive in life, or lack thereof, is dictated by your experiences with your parents. Core memories, those that are often "firsts," remain instructive to our survival for the rest of our lives. If we fall down as a toddler and Dad says "Brush it off," in that one moment, he has welded an instinctual reaction into our true nature. With facial expressions and tone he has given us a rule to follow for strength and composure that we will always abide by. The long-form lesson: we have to repress pain and focus on the task at hand. But in short form, it's simply "Brush it off, don't cry."

If a parent can teach us something that might be helpful, they can teach us subjectively unhelpful things too. If we fail a test fair and square and our mom says "That teacher is too strict," we may adhere to the potentially debilitating habit of making excuses for our failures into adulthood. The results are as plentiful as the amount of people walking this earth. Since none of us are perfect beings, none of our parents are perfect parents. We take from them important tools that form a way of life that serves us well "until," as Kate so insightfully says, "it doesn't."

My dad was hardly a police officer. In his younger adult life, as a moonshiner and cockfighter, he often found himself juxtaposed entirely to the principle of law and order. Still, I think he had more than a few things in common with Kate's dad. They both taught their kids to think critically before they act.

I can't imagine having a police officer for a dad, much less a detective. Not just the pressure to never get into trouble, but knowing he would find out anything I did: those things would have certainly changed my high school experience. From covering our coach's house in toilet paper to doing donuts with a car in the opposing coach's yard, to sneaking girls out of their bedroom windows late at night to steal a kiss, as strict as my dad was, he didn't have a sixth sense of my mischievous plans. Kate grew up differently than most of us, but her experience seems to have served her well.

Something both our dads taught us, for better or

worse, was to not let something like emotions stop us from accomplishing the task at hand. Or perhaps less productively, they taught us not to let emotions let us look weak. When Kate says her dad taught her to control her emotions, or to at times shut them off completely, that's called compartmentalizing. It is a coping mechanism humans use to essentially ignore the psychological fallout that comes with grief, remorse, regret, pain, or sadness. It's a tool used to prevent the debilitating state brought on by bad news or negative experiences. Whether this learned trait is a good reaction or a bad one is a question for more educated mental health professionals to debate, but I can tell you from firsthand experience that it is about the only way to get through a rough deployment, or even an injury like losing your legs, or doing CPR on your own dad as he slips away from mortal life.

I suppose everything we gain does come at a cost, and suppressing those emotions in the moment is a necessity to get through a crisis. But it doesn't make them go away. Like my papaw would say, "Son, scaring the roosters away now does not stop them from coming home to roost tonight."

Just as Kate alluded to, we eventually run out of space in the giant old suitcase in the attic of our brains, where we store painful things and must eventually embark upon the dangerous task of sorting through our baggage.

★★

A SACRIFICE LEADS TO A CALLING

I graduated from the University of Massachusetts Amherst with a degree in psychology. At one point, after working at a horse rescue operation, I had considered zoology—a field that would have fed my love of animals. I settled on psychology thinking that understanding people would be useful no matter what career path I pursued. I thought that becoming a therapist would be a lot of fun. I enjoy trying to figure people out. I guess that's one of the things I learned from my dad. As a homicide detective, he described a murder as a puzzle that has to be figured out.

I saw therapy as being like that: discovering all the pieces and how they go together. But becoming a therapist takes a lot of training. I wasn't going to spend that much more time in school.

My early fascination with becoming a homicide detective had faded as I got older. I got a job working for the Massachusetts Division of Marine Fisheries. I loved the work because I basically got to go boating for a job. I was always curious, always wanting to learn and experience more, so when that job wasn't fulfilling, I hoped to continue being around water and applied to the United States Coast Guard in 2016.

I had a close relationship with my grandparents, but not as close as my older brother, John Robert, who spent a lot of time with them in Florida while a teen. Eventually he applied to the Hillsborough County Sheriff's Office, joining as a deputy. Though he lived out of state, he kept in touch with family back on Cape Cod, calling as many as two, three, and four times a day to check on them.

He was also an accident investigator, a role that involved visiting people after they'd been in wrecks to get a statement. On March 12, 2016, he was working in that capacity. He was on his way home after having visited two different hospitals to visit individuals who had been involved in traffic accidents. It was around 3 a.m.

While traveling eastbound on the Selmon Expressway, he spotted way ahead a drunk driver traveling westbound in his lane. The driver of the vehicle just ahead of him was flashing her headlights, hoping the driver would slow down. Recognizing what was about to happen,

John passed her and placed his vehicle in the path of the oncoming car. This put him in the direct line of the wrong-way driver. The head-on collision that resulted killed both my brother and the drunk driver. The other driver who witnessed it said she heard his engine accelerate as he went past her and believes he intentionally put his vehicle in between them, saving her and her passenger's life, since it would have been them who collided with the drunk driver instead.

Obviously, there is no way to know what he was thinking, if he did it to save them or if he thought he had enough time to get in front of them and try to stop the car. I myself was in a situation at work where a vehicle was going in the wrong direction and heading straight at me. At first you question what you are seeing, and then you have a split second to make a decision: Do you put your lights and siren on and continue ahead, praying they stop their car? Do you try to get in front of the other vehicles to stop them to keep them from being hit? Do you swerve out of the way and let the vehicles behind you fend for themselves, hoping they swerve as well? Thankfully in my case, the other driver realized their mistake and quickly turned around. I imagine some of these same thoughts went through his head.

I remember getting that early-morning phone call from my grandfather. Waking me, he asked me to give my phone to Dad. He added that I shouldn't tell Mom that he was calling. I didn't know what was wrong, but I knew that something was.

My dad walked off, and then I saw him drop to his knees. Honestly, I think that was the first time I ever saw him cry. My mom took the phone, because Dad had just shut down. My mom wanted to know everything. After she got the bad news, she spoke with detectives down there. She wanted to know every detail. She had to know *exactly* what happened to John Robert. It was interesting to see how differently they grieved. My mom and dad reversed roles then. My

I also remembered being prescribed a medicine to take wi．middle school. Try as I might, I couldn't get the things down. J Robert spent hours with me trying to get me comfortable with takii． a pill.

He wasn't like that with his brothers so much, but he definitely was always on the watch for me. Like I said when I spoke at his memorial service, "He was a protector long before he wore a badge. We lost a guardian angel down on earth, but he gained one up in heaven."

If I resisted John Robert's influence in any way, it was this: he had urged me to come down to Florida and apply to become a Hillsborough County deputy and work with him.

"There's no way I'm becoming a cop," I told him. "There's no way I'm coming to Florida. That's never going to happen."

At a reception after the first of two memorial services held to honor John Robert—the second on the Cape, the first one in Florida—three of John Robert's colleagues urged me and my brothers Pat and Mike to do what John Robert had encouraged: join up. While in Florida for the memorial, I decided to take them up on their offer to do a ride-along, just to see. Just to experience a little bit of what John Robert's life had been like—what he so loved doing.

The guy I rode with really loved his job. I fell in love with it too. The squad I was with told me that he loved getting up in the morning and going to work. He was happy. I had never really felt that strongly about any of the work I'd been doing. I could see that this was the kind of work that really mattered. The day after the ride-along, I submitted my application. By October I was in the academy. Becoming a deputy where John Robert had worked was a way of keeping his legacy alive, keeping his name known down here. It was almost like I was picking up where he left off and doing the things that he couldn't do anymore.

dad's emotionless responses didn't kick in. He was a cop. He had seen plenty of bad things happen, but this was different.

I was also put in a situation that called on me to do something I normally loathed: speaking publicly. A law enforcement officer losing his life in the line of duty, particularly under the circumstances John Robert did, is newsworthy. He was being called a hero (justifiably) for having sacrificed his life to protect that other driver. Given the family's history of service, the story was considered even more poignant. Media members called to speak with members of the family. John and Terry flew to Florida immediately. Someone had to step up, and I did so in the immediate aftermath of my brother's death.

I didn't really have time to think or grieve that morning. We got the call at 5 a.m. from Bump, as we called my grandfather, and by 8 a.m. the media was calling. I honestly can't tell you what I said, but it's almost like I shoved my emotions back down, realized that I had a job to do, that I had to handle it, and did. What really mattered was that we got John Robert's story out. People needed to know the kind of guy he was. How selfless he was. How kind. Later, when I delivered his eulogy, I told people how having a big brother like John Robert was the best thing.

He always helped me to feel safe.

I was especially close with John Robert. Pat and Mike were a pair, and John Robert and I were as well. He was five years older, and he'd watch me when I was a baby. One time, Mom had fallen asleep and John Robert peeked into my crib. My ears were pierced, and he noticed that one of the earrings I had been wearing was no longer in one of my lobes. He hopped on the phone to Grandma, worried that I had swallowed it, and not wanting to wake our exhausted mother. He was ultimately able to wake my mother up after my grandmother told him to, and everything was fine—the earring was located.

Hearing Kate tell how her parents reacted to the shattering news of losing a child really shows how humbling and life-shattering trauma can be, no matter how prepared we may think we are. Not that one of their reactions was better than the other in any way, but there is something raw and enlightening seeing that her dad built so much of his life and ethos around emotional control, but when it was one of the things he loved most, well, does anyone really have "control"? I don't think I would.

It seems for Kate, the reaction was responsibility. In the Marine Corps, we had a saying often proclaimed to those not yet baptized in blood and bullets, in combat: "You don't really know what you are until you are forced to find out." I have often parroted this myself post-injury as a means of reassuring folks they could get through it too, when they pour a little too much adulation on me for recovering from my injuries.

Kate's brother, John Robert, was forced to find out exactly what he was: a selfless hero willing to sacrifice his own life to save a stranger. But for Kate, the moment of "finding out what you're made of" came when she held the responsibility of honoring her brother's sacrifice. For so many of us, losing a loved one to a specific thing might move us, with good reason, to avoid that thing. If they drowned, we might stop swimming. If they died while doing a dangerous activity, we might not partake in that activity again. But not everyone is wired the same, and Kate's entire life was preparing her to make a decision that might shock an onlooker like us but made perfect sense to her. When her brother died protecting others, she decided to pick up the shield and carry on the mission for him, for their family, for herself. I don't know what compels a

> person to believe all of their tomorrows are worth sacrificing for someone else's today, but Kate had a front-row view of what that cost is and then decided to sign up for it herself.

LIVING FOR JOHN ROBERT

Living a life of tribute is honorable and hard. So is the law enforcement work.

We've had a ton of murders down here in the last couple of years. It's also just been crazy with domestic violence incidents. Those can be really frustrating, especially because you end up going back to the same addresses over and over again. You never get the full story from the people involved either. They give you a half-truth. But the calls that frustrate me the most are child custody disputes. The parents are no longer together, but they have a child in common. The adults are so hateful toward one another that they will do the most vindictive stuff. What I especially have trouble with is when they put all the blame on the other parent and say the worst things about the other person in front of the kids. We get called out to defuse the situation. And, like I said, we're going to the same places over and over again. Don't they see how traumatic that is for their kid? To see their parents fighting like that? I tell them that they need to be the adults in that situation. We get called out there and have to be the adults for them.

After graduating from the academy, I served in a patrol car accompanied by a veteran officer before I was "cut loose"—working independently on my own patrol.

I remember going out on a domestic call in my earlier days. I had been on my own for only about six weeks. This guy got woken up by his girlfriend. She was going through his phone and found a text from another female. She confronted him about it. He punched her in the nose and broke it pretty bad. By the time we got there, she had left the house and gone to the neighbors. She was bleeding all over the place. The guy followed her, leaving behind their two kids, one of whom was autistic. They were teenagers, so their parents weren't young themselves. We ended up arresting the guy. But the girlfriend's defense attorney made it sound like she instigated the whole thing. We did all the paperwork but eventually the charges were dismissed.

★★★★★★★★★★★★★★★★★★★★★★★★★★★★★★★

Katelyn's recounting of the story was matter-of-fact, but I could tell that she was angry at this miscarriage of justice. Law enforcement's hands were essentially tied. She wasn't the only one who was angry about the outcome.

★★★★★★★★★★★★★★★★★★★★★★★★★★★★★★★

I remember one of the kids talking to the state's attorney. He couldn't understand why the charges were dropped. He was like, "He hurt her. Why isn't he going to jail?"

That was difficult for me to accept, so imagine how hard it was for the child to see his mother being abused and there being no legal repercussions. It's discouraging for the people who have to intervene and investigate these kinds of matters. While these domestic calls are the exception, they are a large part of the work that we do.

As much as I enjoyed the homicide cases I watched on television, that reality-versus-unreality dichotomy was a regular part of our investigations as well.

I remember this other time, these two teenagers decided to rob their drug dealer. They broke into his apartment with two guns that weren't loaded. The drug dealer ended up shooting both of them. One died on the scene. When we got there, coagulated blood was dripping out of his brain, forming a pool of it around his body. The other kid ended up dying later. That's the type of stuff we see. Also, it's not just investigating the crime; we also have to do the death notification. We're the ones who have to tell parents that their kids are never coming home.

Doing that is never easy, and I know what it is like to be on the receiving end of such a call.

If it's an eighty-five-year-old grandpa who dies in poor health and has had a bunch of medical stuff, that's not as hard. It was expected. It's the unexpected that you have a harder time with. It isn't about the dead bodies, it's about the living ones, having to see them struggle and grieving. Those are hard for me. I went on a call. A mom had fallen asleep with her six-week-old baby in bed with her. She rolled over on her child and accidentally suffocated it. I think I was the third one on-scene. The EMTs were doing CPR, but we all knew it was hopeless. The baby was blue; it was beyond help, but the mother didn't know that yet. Then, a few weeks later, the same thing happened, only this time it was a grandma with a six-month-old child.

Those cases were heartbreaking. Motor vehicle accident scenes can be horrific too. Just last month, my squad and I responded to one of our detention deputies who was involved in a traffic fatality. He was in full uniform, on his way home after working the night shift. He had barely drifted over the line and hit a semitruck head-on. He was pronounced dead at the scene. It was hard having to see a fellow officer in uniform in such a condition. I imagine it's similar to what the deputies who responded to my brother's crash felt like.

You have to turn down your emotions to do your job; you can't turn *off* those emotions entirely. You still need to be human. And you have to be able to complete that call for service.

When I take a moment to imagine myself in Kate's shoes, I'm almost paralyzed, because I can't understand how she and so many others can do the job of responding to people being absolutely horrible to one another. The notion that what is just or unjust, lawful or unlawful might seem to be in opposition to what we know to be right or wrong, good or bad, fair or unfair. The truth is, not many of us wouldn't want to throw that man in jail, or worse, for beating his girlfriend in front of their kids. But we live in a country of laws, and Kate's job is to not only serve and protect us as individuals, but also to ensure that we abide by these laws. The difficulty for police officers goes so far beyond making decisions on their own actions, they have to concede to the wishes of victims and the technicalities of guilt and innocence in the eyes of the law.

Far beyond the frustrations of dealing with adult victims who are unwilling to seek justice for themselves comes the heartache of responding to child victims. The hardest days of my two deployments to Iraq and Afghanistan involved children. I don't know how those like Kate can do it every day, in their own community. But they do. And they don't quit on us; they do the hard work of keeping us safe and responding in our times of need, all while processing their own empathy, frustrations, and trauma. But their determination, or, as Kate explained earlier, the ability to compartmentalize, doesn't mean they all "get along just fine." When it comes to members of law enforcement, we often expect them to have a superhuman ability to ignore their emotions, or perhaps many of us don't even think about what they are dealing with at all. But the human experience is one of trauma, and is best lived through self-awareness and support.

DEALING WITH TRAUMA

One of the things that our office now does—and I think this is great—is use a system called Critical Incident Stress Management (CISM). So, when we have a big incident like those two calls for service with those two smothered children, they use CISM to help us deal with the trauma. We are usually given three to five days or so to process what we've seen and experienced. Then they hold a peer support group meeting where you can talk with other deputies about what has gone down. Everyone who was involved in the call attends. Sometimes they bring on the staff therapist.

I was pleased and surprised that at one of the CISM sessions, two fellow officers who were at the scene of those infant deaths did speak up. They were ex-military, seemingly the kind of guys who would never speak up and share their emotions. I like the fact that my department is acknowledging how crucial mental health is to the welfare of its officers and others.

Because officers encounter so many different scenarios and come face-to-face with members of the public at times of high stress, the work can be exhausting. It helps that we have one another to rely on. The commonality of our experience produces a kind of kinship. In my case, that kinship is literal. I have a father and an older brother who work in law enforcement. I can speak with them and there's an unstated sense that we "get" one another. I also became romantically involved with a fellow academy student.

But even with an occupation in common, the negatives sometimes outweigh the benefits of having shared difficult experiences. After six years, we ended the relationship.

I don't think I ever really had a full heart to give him. I was still broken from losing my brother. It took a while, but that loss caught up with me. I'm really good at avoiding problems, so if there was something wrong, I could just pretend there was nothing wrong. I was wrong, though. This job will harden you, and it did that to me and it did that to him. When we met at the academy, we were a lot more bubbly. Then, once you get out on the road, you see a lot of jacked-up stuff. That has an effect on you. It's a lot harder to trust people because we tend to see the worst of the worst. It's not often that someone invites us to a birthday party to thank us.

★★★★★★★★★★★★★★★★★★★★★★★★★★★★★★★★★

> As a veteran, I'm well acquainted with the question, What is the cost of freedom? Easy. It's the life of a friend on the battlefield. But it's also the peace of mind for those of us who served with them. But battlefields seem to always be in far-off places and our enemies aren't always simply trying to destroy us. The price of freedom perhaps then is much more local, more day-to-day, more about how we live our lives than who might attack us from afar. Perhaps freedom then is safety, security, and the peace of mind of knowing that when something goes wrong while we are out living our lives freely, someone will be there to risk it all to help.
>
> The price those people pay is quite astonishing when you hear it from someone like Kate. She doesn't get to just have a bad day, or even a rough deployment. For people in her profession, every day is a bad day in some respect. Every day is dealing with people who have shown the worst of themselves, or have experienced the worst a person can. Most "good" people she interacts with are either defensive, because seeing her means they got caught breaking a law, like speeding, or are distraught because they've been a victim of a crime themselves. The sight of blue lights and a badge in those moments aren't exactly a welcome relief or

> moment of joy. The fact is, this kind of work isn't something you can leave in the locker on the way out the door. It doesn't just affect you internally; it affects your relationships too.

FRIENDSHIPS AND REGRETS

Abby was the best. She was good at her job. She had ambition. She was a lot like me and we shared the same sense of humor. I took her under my wing a bit since I was more senior. I knew that I didn't have to worry about her because I knew she was going to handle her stuff. I remember there was one call she went to. My radio had switched over without me realizing it, so I wasn't aware of the call until later. She had called for backup. The call was a domestic, and she was on-scene with a drunk woman who was getting belligerent. She ended up punching Abby. By the time I got the call, I was a few minutes behind. I showed up and the drunk woman was in the back of the car, cuffed and detained. I took a look at her and could see a bruise starting to darken around her eye. I went up to Abby and told her that I had never been more proud of her. She had managed to handle the situation and get that out-of-control woman under control all by herself.

Abby was like a rock for me. I could call her and talk with her about anything, and it didn't matter. So, it's something I struggle with now. Not having her anymore. Not having that best friend to talk to. It's a lot more lonely without her.

After a year of working the same district with Abby, I decided to join the motorcycle unit. As a result, I was assigned to a different district after completing motor patrol officer training. The two of us kept in touch, but different schedules made it hard. We would occasionally find time for walks together with our dogs.

I was looking forward to a cruise with Abby and our boyfriends and others from their social circle. We'd booked a cruise, but when outbreaks of Covid recurred, we changed our plans. Nobody wanted to go on a cruise and be masked up. I and my then boyfriend decided to fly to Yellowstone National Park. The rest didn't want to make that trek and settled on renting a place in St. Augustine, Florida.

I was getting on the plane when I got a message from one of the guys I used to work for. Through his wife he sent me a text saying, "Sorry for your loss." I started taking inventory. My dad had flown down to watch my dog for us and had just dropped us off at the airport. I have trackers on my phone, and I saw that my mom was home. *Can't be her.* I checked my brother and his girlfriend. *They're both okay.* But my mind had immediately gone to who died. Even before this, after John Robert was killed, anytime I got a call at an odd hour, that is immediately where my brain went. You investigate so many accidents, see so much bad stuff, that you develop that reflex. I guess that's a way to prepare yourself. You don't want to be taken by surprise like I had been when we got the call about John Robert.

So, my phone rang and this time it was my former zone partner telling me that Abby was dead. They had just found out at their roll call, and he wanted me to know. It turned out that while they were on that vacation in St. Augustine, her boyfriend had shot her and killed himself. He was a detective on our GRIT squad—Gun Response Incident Team. I knew him, but we weren't close.

The loss hit me hard, as did survivor's guilt and my sense that if I and my now ex-boyfriend had joined the rest of the group in St. Augustine, things would have ended differently. There was no sign that he was capable of doing something like that. She had never mentioned anything to me about him being violent. But my ex was the one who knew him best, and I can't help but wonder, and think that if we'd both been there, they would have had someone to talk with. I think that Abby was going to break up with him, and it's likely that this was one of those if-I-can't-have-you-nobody-can kind of things. But it's hard to say. Would he have done the same thing if we all hadn't been going on vacation at the same time? Was he always thinking that he would do this? There's just really no way to know.

★★★★★★★★★★★★★★★★★★★★★★★★★★★★★★★★

The earth-shattering news of losing a family member, best friend, or coworker is incredibly difficult. When you lose them in the line of duty, like when Kate's brother died saving a life, at least you have the heroism of their final actions to lean on. Survivor's guilt weighs heavy on so many of us who lose someone in an incident we're involved in. But with Abby, Kate is dealing with the struggle of "what if" and "why" at an extreme level. The truth, I've learned, is that we have to accept that we don't get to control everything or everyone. We certainly have an impact on the people and situations in our lives, but the absence of our impact in a negative situation doesn't equate to our having a negative impact. In other words, there's no way of knowing what would have happened had Kate and her boyfriend been there. There's just as good a chance they could have been hurt or worse and even more people would be left to mourn a loss.

The fact is, life isn't fair in any measurable way. The terrible things we endure often come without rhyme or reason. Abby deserved so much more out of life and Kate certainly has carried more than her share of heartache. The fact that Abby's murderer was also someone who wore a badge shouldn't leave a stain on anyone else wearing one. The true heroism of people like Abby and Kate and John Robert and millions of others isn't the badge they wear or even the job they chose; it's the decisions they

> make and why. Heroes sacrifice their own livelihood for the betterment of others. Heroes are selfless in their endeavors and quiet in their complaints. Heroes put others first and are willing to carry this burden for as long as is necessary.

★★★★★★★★★★★★★★★★★★★★★★★★★★★★★★★★★★★★

RIDING THE "IRON HORSE"

I'm just one of thousands and thousands of law enforcement officers who serve. My circumstances are somewhat unique in having become a sheriff's deputy because of losing my brother. And, as you've seen, the losses mounted after that. When I switched to riding the "iron horse," as I like to put it, I was in District Four, the same one in which John Robert served. I was surrounded by people who knew him, who worked with him, and who would share stories with me and I with them. My riding partner for a year was Bobby Howard. When I came down to Florida for my brother's funeral in 2016, Bobby rode escort so that the family could view John Robert's body. Later he was among four other motorcycle officers who rode in the funeral cortege.

They were total strangers then, but they came up to me and told me that they hoped I knew I still had brothers down there. Then I was partnered with Bobby.

Sadly, Bobby lost his life while riding a motorcycle, but not while on duty. He left the motor patrol squad out of concerns for his safety.

I lost him too. It's been a struggle with the sheriff's office. But the way I look at it, I know John Robert, Abby, and Bobby wouldn't want me to be miserable or sad and lonely and upset and throw my life away. They would want me to keep moving forward and being the best I can be. So that's the way I honor them: by making the best of my life and doing the things they can't do. So, if they're looking down, they'll be proud to say, "That's my girl down there."

I met Kate while she was on duty. I was in her town speaking at a fundraiser, and the event organizers had asked her to come in an official capacity to be security for me. In uniform and with a smile she greeted me without sparking a conversation. She was equally as professional in her responsibility to stand in the background with an eye on the crowd as she was an absolute pleasure to get to know as I forced a conversation upon her. By the end of the night, I knew I had met someone special, someone with a story to tell, someone who might inspire others in their own trauma, heartache, or adversity. I didn't get a chance to meet the heroes in her life who have guided her on this journey, but I believe I can say without a doubt that they are all proud of her. Not just for the decision she made to become an officer, but for the courage she has to recognize the importance of addressing trauma and now using her own trauma to help others.

The Confidence of Camaraderie

TOMMY WEHRLE
★ **SWAT Sniper** ★

"Some days, this job will get the best of you. Some days, this job is the best."

WHAT MAKES A SWAT TEAM MEMBER

Guys come from all walks of life to do this job. The job of a SWAT officer is certainly a unique one. I often think about this when I'm on a barricade, deployed as a sniper. The world slows down when you're viewing it through an 18x precision rifle scope. Sometimes I wonder where the time has gone, and how I got here. How in the world did I end up being one of the guys who got called here to handle *this* situation, and lie behind this rifle?

I knew from the start that I wanted to become a SWAT sniper. Growing up, hunting, and being in the woods nonstop planted the seed for my career path. Learning how to shoot long-range at an early age helped me tremendously when it came time for Sniper School. Crawling around in the woods and shooting shit was something my family did for fun, I never would have thought you could make decent money doing it.

There are many different avenues that you can find yourself in while working in law enforcement. You can be a detective in robbery, burglary, homicide, sex crimes, or child advocacy. You can be promoted and eventually become a supervisor. Everyone has their own personal reasons for doing this. Sometimes the patrol guys talk shit about us because we're supposedly "unapproachable" and egotistical. They call us prima donnas. There may be something to that, but SWAT isn't unapproachable. We're knowledgeable. If you approach us for help, you'll get plenty. We're not unapproachable. SWAT guys *do* love to talk shop, especially in front of people who *aren't* SWAT. Is it cocky? Or just confident? It's probably a bit of both, but I'll go with the latter.

The camaraderie gets tighter the more specialized the field. I remember thinking the camaraderie was really something among Eagle Scouts when I was sixteen. Then I joined the volunteer fire service and saw another level of commitment. Then I went into law enforcement and was selected to go to SWAT. The level of implicit trust and brotherhood in this unit is something I know I'll never experience again.

Either way, who doesn't like to occasionally mention how we "worked real hard" this week, meaning shooting thousands of *free* rounds of ammunition at the range, running CQB (close-quarters battle) drills using NODs (night optical devices), climbing a thirty-foot caving ladder to board a vessel, working land navigation, or maybe just rappelling out of a helicopter? The list goes on. For a law enforcement officer, this job is the shit, *but everything comes at a cost.*

★★★★★★★★★★★★★★★★★★★★★★★★★★★★★★★★★

I met Tommy on a bear hunt in northern Maine. We share a mutual friend, Dave Hentosh, who lives up there and guides bear and moose hunts, primarily for military and first responders. I met Dave a few years earlier on a moose hunt for our *Fox Nation Outdoors* show. Jeremy Judd, a Maine game warden who's also in this book, worked with Dave on that hunt as well.

The purpose of these hunts is to get men and women in these dangerous and high-stress jobs out in the woods around a campfire and talking to one another. That's how I got to know Tommy. He didn't just sit and tell me one heroic story after another; he listened, showed respect and homage to the other military veterans and firefighters I'd brought with me (Keith Dempsey and Clay Headrick were two of them), then politely joined the conversation, humbly adding some of his own experiences to the topic we were already discussing. His humility, coupled with his desire to be of service to others, really impressed me. As the hunt wrapped up, I could tell he was going to be one of us, a part of this brotherhood of hunters and servants that stay in touch via group texts and offshoot one-on-one conversations.

It's funny that he started this conversation off by admitting the bravado often attributed to and commonly accepted by the SWAT community, because I really didn't see that in him at all. Well, until we went to sight our guns in. Then he went full nerd on marksmanship. It was funny but also showed just how passionate and well-studied he is in his craft, the art of neutralizing a target at great distances. The job of a SWAT sniper.

The cost Tommy refers to when accepting the responsibility of his job can be paid in full, all at once, such as being severely injured or killed in the line of duty. Or it can come in increments over a long and tumultuous career, such as seeing innocent civilians or even fellow officers severely injured or killed. One of these incidents that Tommy experienced has become fairly well-known in his jurisdiction and in law enforcement departments around the country.

"In Harford County, everybody remembers what we refer to as the 'Panera Bread Incident,'" is how he began the story. On February 9, 2016, Harford County sheriff's deputy Patrick Dailey was dispatched to a Panera Bread restaurant in Abingdon, Maryland, about twenty miles north of Baltimore. The sheriff's office had received a call about a wanted individual being spotted there. Upon entering, Dailey walked up to a man who was sitting by himself at a table. Their contact was brief. Without warning, the man, sixty-eight-year-old David Evans, pulled out a handgun and shot Dailey in the head, killing the officer.

Evans then fled out the back door of the location. That's when Tommy and his SWAT team got the call. Another deputy, Mark Logsdon, rushed to respond around the same time.

THE PANERA BREAD INCIDENT

We heard on the radio that an officer was down, so the SWAT team was called in immediately. I was in the office planning an operation for the following morning. Luckily, we had the radio on Southern Channel and heard the screaming. Pat Dailey had been shot in the head, and the suspect fled the scene. Everyone responded. While driving there, I was wondering what I would encounter when I arrived. What would I see? What would Pat look like? Although it was against policy and extremely unsafe, I somehow managed to call my wife and tell her what had happened and let her know that I'd be out for hours, if not days, looking for this asshole.

Moments later, I arrived and parked next to another operator, Anthony DeMarnio. We call him "D." We were the first team guys to arrive, along with Nate Gerres. We all thought the suspect fled into a heavily wooded area behind a retirement home. We were preparing for a woodland search when Nate suggested that we clear the cars on the lot first.

This proved to be a pivotal decision.

As we started looking in and under cars on the lot, we heard yelling and gunfire about fifty feet behind us. Deputy Mark Logsdon and the suspect were now in a gunfight. The suspect had been hiding in the front seat of a small blue car in a corner of the lot. As the exchange of gunfire was occurring, I went on autopilot. I viewed this as an "active assailant" incident and ran to the driver's side of the car, near the A-pillar.

As I moved into position, I saw the suspect, leaning way back in the driver's seat, trying to conceal himself. He also appeared to be still looking around for Mark. In nanoseconds, I somehow saw his silver handgun and snapped up my M4 and went to work. During the engagement, I never saw D, who was standing right next to me, also firing. According to the tapes, we fired thirty-three rounds in less than 3.5 seconds. Interestingly enough, I only had a recollection of firing six or maybe eight rounds, tops. After numerous police-involved shootings, this one was responsible for the majority of my hearing loss.

We also didn't forget that with the suspect dead, we needed to render aid to the other officer to try to save him. D worked on him for a while. Despite our best efforts, he ended up dying. We had the funerals.

The shooter was a fugitive from justice. Fifteen years before, he took a .22-caliber revolver and shot his wife in the neck. She survived. He fled to Florida and was on the run for fifteen years. The weird thing is, there was no warrant out on him for the attempted murder. He was a suspect but there wasn't enough evidence for the grand jury to issue an arrest warrant. He did have two other warrants out, for a traffic offense and some other minor matter.

He came back up to Maryland to the neighborhood where he once lived with his wife, and where his family suspected he had been stalking them and frequenting the Panera Bread. It turns out that he was sitting in that same seat in the back for the last two weeks—every day. The staff there thought he was homeless and were feeding him.

The investigation that followed revealed that he had numerous rifles, suppressors, and handguns and over 2,700 rounds of ammunition. He was hell-bent on killing her. Pat approached him without backup asking for identification, and in less than nine seconds after Pat came in the restaurant, he was dead. The shooter stood up, calm as could be, and walked out of there, while everyone else was screaming, running, and turning over tables to get out.

> ★★★★★★★★★★★★★★★★★★★★★★★★★★★★★★
> Tommy's matter-of-fact description of that incident is indicative of how so many in his profession process it internally. We may water down or inject "bedside manner" into telling a story to mixed company. But at night, when we lie down, it isn't the PBS version of the events of that day that we dream about. Unfortunately, that doesn't leave much opportunity to decompress or address the traumatic nature of what he and others experience.
> ★★★★★★★★★★★★★★★★★★★★★★★★★★★★★★

TAMPING DOWN THE TRAUMA

I call it the "trash compactor theory." We see so much horrible shit and we just keep cramming it down and pretending like it's not there. Emotional breakdowns can happen at home when things get to be too much, but they absolutely can't be allowed to happen at work. This job takes its toll. It makes you basically look at everyone like they're a piece of shit until you verify otherwise. This is what keeps us safe. The guys on the team sometimes are all that you have supporting you when you need to decompress. Family can try, but they can never fully understand what we go through on a daily basis. We work with the same group of guys, day in and day out. We see each other more often than we see our families.

Guys come from all types of upbringings. We have country boys and city slickers. Meatheads and skinny dudes. We have young bucks in their twenties and old farts in their forties. Hell, Rip made it to fifty, before he threw in the towel after ten years. We were sent to SWAT selection together. He's still in phenomenal shape.

About half are former military, seeking a similar calling. Kevin, who always tells us he "used to be really fat," now runs marathons, and during PT, his heart rate is lower than yours, *even when he's wearing a gas mask and running next to you.* Marco is the prettiest one we have. The "Italian Stallion," as we refer to him, has spent more than fourteen years on the team. He plays and coaches soccer. We rely on him to get us in the door—he's an Explosive Breacher.

CJ is our FNG. Fucking New Guy. He came to us after eight years in the Army as a forward observer. He doesn't know it yet, but the former soldier is definitely team-leader material and will do great things in this unit, long after I'm gone. He might be short, but you don't want to fight him. He often does Go-Ruck challenges and will go for a twenty-miler, just because.

I also can't forget about Anthony, or D. He's a mountain of a man who's just a big teddy bear. The boys can tell his mood immediately at training the second he walks through the door. He's a jiujitsu brown belt and wears his heart on his sleeve (both literally and figuratively) with his full left arm sleeve painted with dark ink tattoos, mainly capturing family images and, of course, SWAT stuff. D and I will be close friends forever after being involved in that infamous Panera Bread shoot-out together. Our wives are now besties.

Greg's our team leader. We call him Daddy, since he's the father figure and the second-most senior operator in the unit. He's also a soon-to-be jiujitsu black belt and has a VO2 max of 53 (VO2 max is a measurement of how efficiently your body can use oxygen), which is evidently "superior," according to our tactical Garmin watches. He's a human freak of nature. The former three-gun nation champion has won countless national level three-gun, multigun, and long-range sniper competitions. At one point he was shooting between 45 and 47,000 practice rounds a year.

Greg has trained personnel from elite units such as the FBI Hostage Rescue Team (HRT) as well as some boys in North Carolina down at Fort Bragg. For years Greg and I have traveled to various states together, teaching CQB, entry tactics, and shooting techniques to other SWAT teams. Greg is the guy who, no matter what you do, does it *just a little better*. He's like that with everything. It's incredibly frustrating. You could teach him a brand-new skill and in less than twenty minutes, he would have already figured out a way to do it more efficiently. Your only defense is acting like you already knew that, and that his way is old news.

One of my closest friends, Todd, a former Marine scout sniper with multiple kills, told me years ago that every operator has his "last raid" and "last day" on the team. After that, you're simply a "has-been."

After leaving the team, you're allowed to attend the Christmas party of that same year. If you've done more than ten years on the team, we all chip in and buy you a handgun with the team's logo engraved. After that, you're just a police officer who "used to be SWAT." Nobody cares who used to be SWAT. It's said to be a rather depressing feeling. The game is changing so quickly these days, and when new technology is introduced, like drones for instance, tactics change. Again, if you're not on the team when it's implemented, you're a has-been. Becoming a has-been is not that far off for me. June 2025 will somehow mark twenty-five full years of service for me and I'm considering other employment, hopefully still in the same field. I'll probably get into something related to teaching this shit to the new guys.

Come to think of it—and I really don't know why—we've all called Todd "Shit Bird" for well over twenty years. Most of what I know about countersniper operations I learned from Shit Bird. He was also an Army sniper after getting out of the Corps and served as an aerial door gunner on a little bird while contracting. After serving on multiple combat deployments in Iraq and Afghanistan and contracting with Blackwater and pretty much every legitimate entity under the sun, he has found a job in law enforcement and is currently working as his agency's head firearms instructor. Although he's far from it, he often refers to himself as a has-been. We still teach courses together every year. He's forgotten more about tactical operations than I'll ever know. Todd and I still conduct training and teach courses together in the private sector.

Being able to live without being attached to my phone 24/7 and being constantly called out will be a refreshing experience for my family. My wife, Nicole, and our twelve-year-old son, Max, have been dealing with my unpredictable schedule disasters, middle-of-the-night callouts, and my unannounced temper tantrums for well over ten years now. My five-year-old son, Hunter, doesn't understand why I keep zipping up a green flight suit and running out of the house several times each week muttering things like "I can't fucking wait to be done." The truth is, I will miss this line of work immensely.

★★★★★★★★★★★★★★★★★★★★★★★★★★★★★★★

Teamwork is something I know well. Whether it be a SWAT or military unit or a sports team or a sales team, what Tommy is describing is something that is truly primal among human beings. If you watch a herd of deer, specifically a bachelor group of bucks (males), you see that there are alphas, the dominant leaders, and betas, the old guys. The bucks who have fought the fight and learned the hard lessons are too old to challenge the alphas, so they don't partake in the annual ritual of "fighting" to establish dominance. Then you have the omegas, the serviceable young bucks who can hold their own but aren't ready. They primarily take care of one another. They stay to themselves and live out their lives.

Humans aren't all that different. We don't all have the desire or even fundamentals to be the alphas. But we all want to know we have a place and serve a role. We want to have some responsibility, and we want to know there's an ecosystem of others to have our backs. That's the team. That's the primal nature of man. The importance of being on a team doesn't just apply to an occupation; it's how we make sense of the trauma in our lives. It's how we know we aren't alone. It's also the sense of responsibility that motivates us to keep going even when something breaks us down or should get more attention to heal than we feel like we can give.

★★★★★★★★★★★★★★★★★★★★★★★★★★★★★★★★★

HOW TOMMY DISCOVERED THE WORK

The biggest thing that attracted me to this job, ironically, is that I grew up next door to a fire department. My mother still lives there. I'm still a "life member." Ever since I was young, I was obsessed with the fire department. When I was in high school, I attended a magnet program with courses at the Baltimore County Fire Rescue Academy and got all the certifications. I eventually obtained Firefighter I, Firefighter II, EMT-B, and Rescue Technician. Unfortunately, back in 1997 there was a hiring freeze and I found it incredibly difficult to get hired. Now out of high school, I needed a career. A bunch of my buddies told me, "Hey, you're a gun nut. You're into hunting. You ever think about being a police officer, maybe SWAT?"

I hadn't really thought about it. I didn't know much about the process of becoming one. All I really knew was about the fire department. I would come home from school, throw my book bag down, do my homework, and run to the fire house. I didn't realize it, but those classes later on would benefit me going into the law enforcement route, specifically on a SWAT team.

My friends knew me well and my obsession with firearms and hunting was obvious to everybody. So I took my friend's advice. I applied to the Baltimore County Police Department in 1999 and they said, "Well, you've never been in trouble, you've never had a parking ticket, you've never stolen anything, you've never done drugs. Okay, you're hired."

They hired me immediately, and before I knew it, life took a turn, and now I'm quickly heading down a law enforcement route instead of a fire-and-rescue route. I suppose that was my true calling.

When at the police academy, I saw a guy wearing a ghillie suit and carrying a scoped rifle. I knew immediately what his job was. I thought, *Who the heck is that guy and what do I have to do in order to get his job?* I was told that he was a sniper on the SWAT team. I said to the instructors, "Whatever I have to do to be able to do that job right there, then I'm going to do it."

They laughed and said, "Good luck."

Tommy spent the first eight years of his law enforcement career in the county where he was born, Baltimore County. This should not to be confused with Baltimore City Police, which has about 2,000 officers whose sole jurisdiction is the city itself. The Baltimore County Police also has about 2,000 officers. After that initial eight-year stint, in 2008, Tommy did a lateral transfer to the Harford County Sheriff's Office, one county to the east. He wanted to transfer to pursue a full-time opportunity as a SWAT operator. He worked patrol for about two and a half years before joining special operations in 2011.

LEARNING TO HANDLE GUNS AS A KID

Much of the appeal to being on a SWAT team came from my early experiences hunting. I like nothing better than being in the woods with a rifle, wearing camouflage. It's relaxing. That's where I want to be. Being in the woods is how I personally connect with my true self, the way my father did. Sometimes people joke that I hunt people at night and hunt animals during the day. Although I laugh about that, I take the job extremely seriously and understand the enormous responsibility it takes to be on this team and do this job. The team has a 100 percent mission success mandate. Standards and expectations are high. They have to be. Still, this is kind of what I was born to do. This is my calling.

My dad was the oldest of three brothers. My uncle Bruce is the middle one, and then there's my uncle Keith. I was extremely tight with all three of them, and each had a significant influence on me when it came to hunting, firearms safety, and being in the woods. My dad died last year, but my uncles are still around. Uncle Bruce lived and worked a leased farm. My wife and I bought a place on thirteen acres next to that parcel. We wanted my kids to be fourth-generation hunters on that land.

I took my first deer on that farm when I was fifteen. It was a Wednesday night, and I was twelve. I shot that deer from about twenty yards, and it took off. We had a good blood trail. Unfortunately, it was a school night, and we had been tracking it for a while when the blood trail tapered off. My dad told me that I had to call it a night. We had to head back because "school always took precedence." He told me that he would come back the next day and find that deer, and he did.

I was a little upset that I hadn't been able to get it that night. Either way, that was a milestone in my life, shooting that deer with a bow and arrow. I drive by that tree every day on my way into work. I have so many memories of that farm, and it has had a massive influence on my career. Honestly, I went the way I did in life because of how I was raised around that farm. There is also some irony in hunting. Just as people often misinterpret or fail to fully understand hunters, they often do the same with law enforcement. It's easy to misjudge something unfamiliar to you without having all the facts.

NOT EVERY SWAT TEAM IS CREATED EQUAL

The term *SWAT* certainly gets thrown around a lot in America. It's an acronym for "special weapons and tactics." There are agencies who boast that their SWAT teams are capable of handling critical situations when in reality they are often ill-prepared, ad hoc personnel, through no fault of their own, and don't have even remotely enough time to train all of the necessary skill sets, *which becomes a major liability*. Most of these units are the ones that end up on YouTube for everyone to eye-roll at. These are usually the videos that civilians and military personnel watch, and that's now their impression of a special weapons and tactics team.

However, there are *many* SWAT teams (thankfully, like ours) that are fully backed by their agency commanders, have adequate funding, and employ highly trained and dedicated personnel who have undergone and successfully passed extremely rigorous selection processes. These team members are *fully* capable of responding to and resolving high-risk situations such as active-shooter events, hostage situations, and woodland fugitive apprehensions. These teams can even respond to "tubular assault" emergencies, such as a plane, a train, or even in a boat or a bus. Not too many people realize that there are SWAT teams around that would even rival many military units and surpass them in some skill sets. These are the men I work with. These are the men who will never truly understand how much they mean to me and how much this job has changed me, for better and worse.

One of the skill sets that I specialized in while in the fire service was rappel operations. I was able to take that training and real-world experience and build a Tactical Rappel Master Course, where elite law enforcement and military units can learn how to safely and efficiently facilitate tactical rappel operations for their own teams. We're the only agency in the Baltimore metro area teaching this, and we get guys coming from the Secret Service, the FBI, state and local SWAT teams, and even the military to benefit from what we can teach. I built our program from the ground up, thanks to all the classes I attended years ago.

It's often overlooked, but the main reason why SWAT teams were even formed in this country *was to save lives*. They were created to stop active assailants and conduct hostage rescue missions. As a SWAT team, you always want to facilitate a safe surrender. We go on hundreds of operations and callouts each year. We rarely have to neutralize a suspect, but it does happen. Some of the guys who have been in the unit for a while have had multiple engagements. It's funny how we safely and quietly handle hundreds of missions each year without a shot ever bring fired. No one says *anything* about them. You see, when we show up, we're expected to succeed.

I know there's a big rift right now in America with the militarization of police and law enforcement, specifically with police use of force. The reality is that there are criminals out there who have access to body armor, weapons, and even explosives. There are criminals out there who will shoot at police officers. We have to be prepared to face these challenges.

The SWAT team for my agency, the Harford County Sheriff's Office, is what's referred to as a "full-service" SWAT team, and out of the several hundred sworn, there are only seventeen of us. Our neighboring agency, Baltimore County, has around two thousand officers, but only twenty-four tactical personnel. We support their team on a weekly basis. There's a very rigorous selection process, and there's a very demanding and ongoing training that we have to do. People often think that they've witnessed a "SWAT team" execute a search warrant of a house. This is usually not the case. Just because the police did a "raid" at someone's house doesn't always mean that the tactical team conducted it. Black ski masks and jeans, coupled with an exterior vest and a light on your handgun, doesn't mean you're SWAT. They're more likely narcotics detectives executing a low-level search warrant. On the other hand, when you see two large armored personnel carriers (APCs) roll up with countersniper teams and negotiators, that indicates a true tactical team executing a warrant.

I have been fortunate enough to travel the country teaching other teams, and I realize that in some states, they may train once a month. I don't understand how you can remain proficient in any skill sets when you're faced with so many different environments and circumstances. We have to be prepared to handle pretty much anything. After all, no one else is coming. When police need help, the buck stops with us.

In 2022, Tommy and his team faced a situation that had pretty much everything. A police officer from a neighboring agency had some domestic issues. He was having an affair with another woman, another police officer, and they'd been on a Bonnie-and-Clyde crime spree, including armed carjackings. At one point they went to his house and seized his children from his wife. He went on the run with the kids and his girlfriend. His wife contacted law enforcement and made it clear that she felt certain that he was going to kill his children.

A JOB GONE WRONG

In November 2022, we were contacted by Baltimore County SWAT with a request to assist them with a potential hostage situation. Our team was called out and all operators responded. When we arrived at the command post, we were briefed by a lieutenant who advised us that one of their patrol officers had gone rogue as a result of domestic issues with his wife. The suspect and his girlfriend, also a police officer, had kidnapped his young children and the two had gone on the above-mentioned spree. After initial investigation, it was apparent to the mental health specialists that the suspect would likely kill his children and himself if police tried to intervene. Plainclothes detectives and surveillance videos had verified that the suspect and his girlfriend had been staying at the Comfort Inn on Loch Raven Boulevard in Towson, Maryland, for several days. Additional information suggested that the suspect was an immediate threat to the children and needed to be apprehended as soon as possible. The exact location of the suspect and children was unknown, as they could have been hunkered down in the hotel room, or out driving in the suspects' vehicle. The decision was made to get a signed search warrant for the hotel room.

The mission was to conduct a deliberate hostage rescue utilizing three elements. One comprised Baltimore County SWAT assaulters who would covertly place an explosive charge on the hotel room door. Another element would be on the perimeter in the event that the suspect pulled into the parking lot with the children. My element was tasked with setting up an aerial diversion from the roof. The plan was to initiate the operation by deploying a noise-flash diversionary device (flash-bang) dangling from a rappel line several stories up, outside the suspects hotel room window. This particular flash bang was a nine-banger, meaning it would emit nine loud bangs in succession. While this was occurring, the explosive charge on the door would go off, opening the hotel room for the assaulters to make entry and locate the suspect. We all knew that if we located the suspect in proximity to the children, lethal force should be used since he was considered an immediate threat to them.

After the breach, it took the assault element less than nine seconds to secure the hotel room and verify that no one was in there. Now investigators shifted their attention to attempting to locate the suspects' vehicle, which was likely to contain the children. Neighboring agencies were notified and a look out was sent for the vehicle. Unfortunately, the exact nature of the investigation was not relayed properly to surrounding jurisdictions. We found out a patrol officer located the suspects' vehicle about forty-five minutes west of Baltimore. The officer recognized the vehicles plates and initiated a traffic stop. The suspect was in the back seat with the children and the girlfriend was driving. Instead of pulling over, the vehicle took off and police pursued, not knowing that a hostage situation was unfolding in front of them. The suspect eventually shot the children, shot his girlfriend, and killed himself. The vehicle slowly rolled off the road into a fence and came to a stop.

While our portion of the search warrant was a success, the overall mission was a failure for law enforcement. If only proper communication would have occurred, the appropriate assets could have deployed and executed some other type of plan.

This is one of those unique situations in law enforcement where remote neutralization is likely the only realistic option to save lives. A news headline about police rolling up at a traffic light and executing someone would be a tough one to read. However, traditional methods, such as a felony stop (where we surround but don't attack a car), vehicle jam, or vehicle assault, would never have allowed enough time for operators to approach the car and save those kids. One well-placed round by a highly disciplined SWAT sniper would have likely saved three of the four people who died that day.

Most of us go through life never having to consciously make a life-or-death decision. We actually do make those decisions multiple times a day. When we pull out onto a road, when we see that food has spoiled and throw it away, when we tell our kids not to run with scissors or knives, we make decisions that save a life, or at least prevent death, every day.

But none of those decisions come with the choice to take a life to save another life. That's called lethal force and is the type of decisions police officers like Tommy make seemingly every day. As Tommy points out, many in the public sit idle waiting to have a passionate opinion about a lethal incident. It's hard to put yourself in the officer's shoes because you don't have the experience or the training to know what he or she was aware of or could have reasonably predicted in the moments before employing lethal force. The truth is that sometimes they get it wrong. By accident or confusion they misinterpret a perp's actions or words and have the responsibility to respond. Sometimes they are negligent in their use of force and are convicted of a crime. But most of the time—times we rarely hear about—they are reluctantly using the last justifiable means available to take out a danger and keep innocent lives safe. It's a tough job, not just because the job itself is hard, but because there is no room for error when lives are at stake. And even when you are justified in using lethal force, the emotions you experience are real and raw and stay for a long time.

WHEN LETHAL FORCE IS THE ONLY OPTION

The determination to use lethal force always rests with the tactical police officer who has been deployed on-scene. People ask me all the time just when, as a sniper, I get the "green light." Neither I nor any of my snipers must wait to receive a green light from anyone. If we perceive that a person is a threat to themselves, another operator, or a member of the public, we can immediately engage them. That doesn't mean we won't be held accountable for all the rounds that get fired. It is simply that we do not need to wait for "approval" to shoot someone. The justification rests solely on the operators who make the decision to take their gun safety mechanism off and press the trigger during a critical incident.

Using the hostage situation where the suspect was holding his children hostage in the car, when we look at the totality of the incident and all facts involved, not only would that have been a textbook justified shooting, but we actually could have been held liable if we did not act. That's because, looking at the severity of the overall incident, he was clearly a threat to other people, and he acted on that threat. Tasers, less lethal munitions, negotiations, and all the other law enforcement traditional tactics simply do not apply here. This particular situation could have been solved with a circuitry shot to the cranial vault, stopping any brain activity and not allowing the suspect to take further action. Sometimes you need to take a life in order to save another.

So, that's one that you have to put in the vault, or as I like to say, in the "trash can." You know how it is when the can is getting full and you don't feel like emptying it. So what do you do? Mash down the trash and add more. And as long as the top of the trash looks good, and isn't spilling over, put more in there.

That's exactly what you do. You are seeing things and hearing things and dealing with things and then you come home. My family knows what I do for a living, but they really don't *know* what I do for a living. They certainly don't know what it's like to deal with all that, to see all that stuff. At times it's like I'm trying to act like I'm an active participant in a conversation or situation at home, but my mind is completely somewhere else. That's really hard on everybody involved.

If you don't have a family that supports you, it is impossible to do this job well. I guess if you're twenty-seven and single with no responsibility, maybe, but you really need a healthy support system at home. We have all seen horrible things, and it changes you.

★★★★★★★★★★★★★★★★★★★★★★★★★★★★★★★

The phrase "suffering in silence," used for men like Tommy, only means the person experiencing these things is the one suffering. The truth is the psychological effects of combat or emergency response reverberate throughout their families. Humans are much more intuitive than we give ourselves credit for. We tend to believe that our spoken words are the only ways we communicate, but the truth is that every expression on our face, the tone in our voice, our body language, the things we choose to do, and when and how we choose to do them are all forms of communication. Our families or the ones we live with pick up on this more than anyone. They may not realize it. It may not be a fully conscious processing of information, but the derived feelings and tensions within the home caused by this can be extremely destructive. The suffering is shared, and having one person in a home serving in a job like Tommy's becomes a burden to all. The entire family is giving up some peace for the good of the community their hero serves.

Thankfully this is being talked about more and more. People are starting to realize how this happens and are talking about it. Some folks have even started their own nonprofit organizations to aid in getting these men and women to a better place. Not just for their own mental health but for their entire family. One of these folks, Dave Hentosh, whom I mentioned earlier, created an organization called Veterans Afield Foundation. His simple but impactful mission was to get guys like Tommy and myself together in the natural environment, with fewer distractions and stress, and let us reset through a shared mission (a hunt) and organically talking about our shared stresses and experiences. It's working.

AN EXTRA FAMILY THAT ALWAYS HAS YOUR BACK

The camaraderie with the guys on the team helps us all so much. The relentless ball-busting that goes on is our love language. These guys are good at a lot of things; they're all high achievers . . . and they're also kindergartners. They have the ability to perform at a high level and then descend into childlike antics. They can flip that switch.

That camaraderie extends beyond our team to our wives. They become close, but I still don't think that after all these years, my wife truly understands how tight I am with the other guys. When you work with people who wouldn't hesitate to sacrifice their lives to save you, that's something. And they know that I would do the same thing for them. Together we've put ourselves in harm's way. That's what recharges me. That's why I keep doing this.

There are people outside your family who do help you—other first responders and military veterans. A few years back, I booked a bear-hunting trip in Maine. When I got there, I found out that I had been chosen as the recipient of a moose hunt, specifically for veterans or first responders through the Veterans Afield Foundation. To my surprise, all of the hunters in camp were either veterans or first responders who had been through tough times. I had no idea of the impact that one week would have on my life.

As the week in Maine progressed, I was fortunate enough to harvest a bear and a cow moose. I was grateful for the opportunity to hunt these animals. More importantly, I was humbled by the people that I met and their stories. We'd all been through similar experiences in our careers. The friendships that you make on these hunting trips are critical to your mental survivability. You would absolutely implode if it weren't for unique opportunities like this.

After that hunt, I had a miserable thirteen-hour drive back to Maryland, with plenty of time to think about things. I was flabbergasted at how quickly the men in camp bonded, simply by leaning on their previous experiences to initiate genuine conversation. I have actually done a few events with VAF now and I knew that I wanted to give back. I have since committed to donate proceeds from my own private company to help with events for them. In addition, I donated several shotguns and rifles to VAF so that hunters have their choice of weapons when they come to hunt in Maine. I plan on being as involved as I can. I will never forget the feeling when I walked into camp for the first time. I wasn't quite sure what to expect but was eventually greeted by a few humble boys from Fifth Special Forces Group, some of whom I still stay in contact with today. We have several hunts planned.

Maybe that's the silver lining in this whole thing: the relationships forged, and the hard lessons learned.

★★★★★★★★★★★★★★★★★★★★★★★★★★★★★★

The impact we have on others is often missed on us. We go on about our lives and that moment we shared with someone may change their outlook on things forever, and vice versa. That's how these hunts work so well. Keith, Clay, and I have remained friends with Tommy since that hunt. On the outer level, we simply shared a cabin and a few meals for a few days. We sat in our hunting spots alone and with little cell service, but coming in from those hours of solitude paved the way for impactful and life-changing conversations over those few meals. Tommy will be the first to tell you that such experiences don't make everything go away or make your internal battles fade. But it is important to know that you're not alone. And as rare as men like Tommy are, there are others dealing with similar issues, and it's good to know them and have that lifeline to reach out.

CRIMINALS WITH DEATH WISHES

This call came almost a month to the day after one of my friends, Jason Schneider, was shot and killed while Baltimore County's Tactical Team was executing a search warrant. Jason and I went to the academy together. He had been on their SWAT team for thirteen years. They raided a house, and Jason went in pursuit of a subject who was fleeing. Despite taking the rounds that killed him, Jason returned fire and killed the subject. Two others in the house were arrested on weapons charges. One was found not guilty; the other pled down a five-year sentence to eighteen months. That's how those things go sometimes.

We received a callout for a hostage situation. The suspect had taken his girlfriend hostage and was making threats to harm her. He also stated he had multiple firearms inside the residence. The situation had escalated as a result of a domestic dispute minutes earlier. Tactical personnel arrived on-scene and established a perimeter while negotiators attempted to make contact on the phone. Throughout the night, negotiations deteriorated significantly. Suddenly the suspect came to the window with something in his hand. I could see through my scope that he was carrying a cell phone. I was prepared to engage him, but he was not armed. He left the window and came back a few moments later. He opened the window again and leaned out with another object in his hand. We were still using green and black phosphorous night vision at that time, and I was sure he was brandishing a black semi-automatic handgun. He was waving it around and pointed it right at my assault element. My job as a sniper was to provide operational intelligence and lethal coverage. After watching him point what appeared to be a gun, I took my safety off and fired one round. The suspect immediately dropped inside the window and out of view.

Days later, while being interviewed by Homicide, I learned that no weapons were recovered at the scene. The homicide detectives showed me a photograph of a metal object that the suspect had fashioned to look like a gun, in an apparent attempt to get us to kill him. We found out later that this individual had attempted suicide by cop numerous times in the past.

To this day, when I teach basic and advanced sniper schools, I use that fake-gun incident to demonstrate the point that you have to act based on facts known at the time. You don't get benefit of 20/20 hindsight.

I want people to know that there are guys out there doing operations like that one, and many others. They do this every single day. They're kicking in doors at two, three, four in the morning. They're at high risk, and it's easy for people on the sidelines to second-guess or Monday-morning quarterback an incident.

THE DARK SIDE

I'm constantly on alert. When I'm in a restaurant, I have to take a seat in the corner. I don't go anywhere without a gun on me. I don't like to fly. When I do, I try to sit at the back of the plane and see everybody in front of me. For the last twenty-four and a half years, I've been in a great many high-risk situations, and that makes you see the world differently. And my career affects my wife. She's been with me when I've found out I had coworkers murdered. She's attended police funerals. So it's not just me.

I think back to the day of the Panera Bread shooting and that moment when I learned of Pat being shot. I remember telling my wife that I was infuriated and that I was going to find that motherfucker. That was raw emotion pouring out of me. Six minutes later, D and I were the ones who took him out.

I remember being back in the precinct, in the lunchroom, when we got the word that both Pat and Mark were deceased. I was relieved that at least the shooter was dead. I hope that it gave Pat's and Mark's families some closure in that they did not have to go through the torture of a trial, if you can call that closure. In some small way, I suppose it validated that all our countless hours of intense training were for something worthwhile.

Some days, this job will get the best of you. Some days, this job is the best.

The Catch-22 of Police Work

JUSTIN HEFLIN
★ Sergeant ★

"I don't want to disappoint my family or disrespect my uniform. I always do a check: Would my family approve of my actions right now if they could see me? If the answer is no, then I need to go a different way."

I don't remember exactly when and where I met Justin, but I've known him now for more than a decade. We were Marine EOD techs together, but stationed on different coasts and deployed opposite one another. We may have briefly met at Camp Leatherneck when he was leaving and I was arriving.

I know we met at an annual memorial for fallen EOD techs in 2012 or 2013, but our current friendship really started when a group of our mutual Marine EOD friends started a group chat to keep in touch a few years later. This group chat turned into an incredibly important line of communication for all of us. It inspired my first book, *Unbroken Bonds of Battle*, and has helped each of us through our toughest days. I hate to use these exact words, but it's truly a support group for all of us as we navigate life post–Marine Corps and postwar.

One of the first things that struck me in Justin's texts to the group was how he would talk about his current career, that of an Indiana state trooper. It was obvious how proud he was of the badge and uniform he now wears, and he took the venting and ribbing from those of us who are prone to get tickets pretty well.

But I could also tell it's not an easy or safe job. So when we all attended a reunion pheasant hunt in South Dakota a few years ago, after a long day of hunting I asked Justin to share more about it. What he had to say opened my eyes to just how dangerous and demanding his job truly is. I knew my friend had a story to tell and we all needed to hear it. So I asked him to share it in this book. What I learned about Justin and why he does what he does every day is something I found truly inspiring.

A MILITARY BASE CHILDHOOD

My earliest memories of childhood come from Hawaii. I've got flashes of the other places we lived, but I'm not sure if those are memories or if they are simply me putting a story behind a photograph. But I definitely remember Hawaii. For some reason I couldn't wait to get away from there. It's weird, in retrospect, given how beautiful the place is. I heard others talk about wanting to move on, that they were ready to go somewhere else, as they commonly did after being stationed somewhere for so long. I guess I had adopted that line of thinking.

My dad was stationed at Kaneohe Bay, on the island of Oahu. We were so close to the water that I could look outside my backyard, and past a long row of base houses and the playgrounds and backyards and see the ocean. I didn't appreciate it as a child, but I can understand now how incredible it was to be there. In school, we were taught the native Hawaiian language. To this day, I can still recite the names of colors in Hawaiian. Although I know the words, I'm not sure which word means which color.

Living on different military installations . . . they're like a perfect America. Through a child's eyes, everyone seems to get along. People from different backgrounds all come together for the same goal. It's a very tight-knit version of America. In my early years that was all I knew. I thought that was just how life was. I didn't realize how different some people were until later. How different life could be.

★★★★★★★★★★★★★★★★★★★★★★★★★★★★★★★★★★★★★

A large part of who I am, my identity as a person, is wrapped up in my pride for where I'm from. If you watch me on Fox News, you see me perk up when someone with a similar accent comes on. If we have a cooking segment around Thanksgiving, I'm vehemently advocating the supremacy of the southern staple of "dressing" over its northern counterpart, "stuffing," and if it's college football season, you can bet I'm wearing my UGA power-G lapel pin. Not because that's my alma mater, but because it's my state's football team. Early on, after joining the Marine Corps, I was truly dismayed to meet folks who didn't have that at all. When you ask someone "Where are you from?" and they respond with "All over" or "Well, I graduated from high school in . . ."? Those answers didn't really make sense to me.

Before that, I hadn't met someone who didn't have a hometown and culture to accompany it. In the Marine Corps, those folks were usually what we call "military brats" or children of career active-duty military members who, by the nature of their service, move duty stations every handful of years. But what may seem like a disadvantage to someone like me, so self-admittedly wrapped up in hometown pride, is really something special for those people who grow up this way. It's not that they don't have memories or culture or a place to call "home," it's just that they aren't limited to one monolithic answer for such a question. They grow up experiencing many cultures, climates, locations within the US, and sometimes even other countries. Home for them becomes a synonym for certain people, not a specific place.

Along with the moniker "military brat," those who grow up the children of active-duty Marines often aren't excited about joining the Corps themselves. Perhaps they feel the need to forge their own path, or maybe they feel like they've already put their time in by living on a Marine or Navy base most of their lives. But a few are quite the opposite. They hear the calling to serve as a family legacy and cannot wait to get their opportunity to wear the uniform

> themselves. No matter where a Marine Corps military brat lands on wanting to serve, one thing is consistent: having a Marine parent isn't easy. That leaves a mark for life.

★★★★★★★★★★★★★★★★★★★★★★★★★★★★★★★★★

GROWING UP WITH A MARINE DAD

I grew up the oldest of five, with two brothers and two sisters. We were all pretty tight. Naturally, we fought like brothers and sisters do, but for the most part we were a pretty tight-knit family. My mom and dad made sure of that. And we all get along great to this day.

Being raised by a Marine father, or at least mine, was really interesting at times. He knew no boundaries. He was not shy about being in our business. He wanted to see us in the morning. He wanted to see us after school. He always wanted everybody together. We always ate our meals at the dinner table as a family. He always had to have some form of contact with us and it wasn't until I was in the Marine Corps myself that it dawned on me what he was doing, and I thought: "That son of a bitch was holding formations with us, and I didn't even know it!"

Every weekend, my dad would tease: "Saturday is field day." We would clean the whole house . . . scrub the toilets, sweep and mop the floors, empty rooms and clean them from top to bottom and then put them back together. My dad loved the Marine Corps. He even punished my brother and me with PT and a mock boot camp for getting bad grades, a punishment that went on until new grades were posted.

Dad, Bobby Heflin, was originally an infantryman but later served as an aviation refueler. He also worked part-time with the military police.

The military was important to my dad, but church was just as important to him. We were pretty involved in the church when we were in Hawaii. We attended a very small, tight-knit church multiple times during the week. Between the church and his military peers, it was like a big family, which is typical.

When my dad was getting out of the Marine Corps and we were leaving Hawaii, I remember folks telling him that they were going to miss him. He made sure to use this as an opportunity to let them know they would see him in the future as an officer on the TV show *Cops*. He was serious. We would always watch the new episodes together as a family, wearing our dad's T-shirts, which looked more like dresses than shirts. He was bound and determined to become a police officer.

My dad did just that. He was a man of service, and with that service comes sacrifice of all kinds. When my dad left the Marines, it was pretty tough financially. I remember our family not having much money. We first moved to Georgia when he joined the civilian world, where he became part of a county sheriff's department. We lived in a pretty rough trailer park, far from the base housing we were used to in Hawaii.

This is when the ideal and the real seemed to first butt heads. Like I said, we didn't have much money. And people knew which trailer park we were from. I remember my dad making a comment that he would only get respect from our schoolteachers if he went to the parent-teacher conferences or other school functions wearing his uniform.

THE CATCH-22 OF POLICE LIFE

My parents met in Indiana during their high school years, but we moved to Georgia because we had family there, and my dad had a job opportunity at the sheriff's department.

Going to Georgia, after he got that job, I think that things weren't quite working out the way he and my mom had hoped. My mom felt the pull to go back to her home in Indiana, so that's what we did.

Both Mom and Dad were born and raised in northwest Indiana, near Lake Michigan and the Illinois line, and close to the once-thriving city of Gary, Indiana, with its steel mills. After we moved back to Indiana, my dad wanted to become a cop again, and he tried for years and years. He had joined the Marines right out of high school, so he didn't have a college degree or any other formal education. I think it was a little different getting hired back then in the 1990s without a degree. So he took jobs that he didn't necessarily see himself staying in for the long haul. Both my parents worked as managers at a Burger King for a while. My dad eventually became a nonunion electrician and learned the trade well. He was a great electrician. Still, he wanted to become a cop. I was a freshman or sophomore in high school when he applied to become a marshal in a small town nearby. He got accepted and went to the academy for training.

At about that same time, my parents divorced, after eighteen years of marriage. I was seventeen. Things must not have been very good between them, and I'm sure the strain of my dad being a midnight police officer did not help. Meanwhile, my dad was finding his way on the job. I remember him bragging about getting the first DUI high-arrest award from the county. (That is basically an award for making the most arrests for the offence of driving under the influence.)

Unfortunately, my dad was not very good at the internal politics involved with his work. That wasn't his strong suit. He was more of a tell-you-how-he-felt kind of guy, which doesn't always go over well with leadership. He left that department and went to work for another one. While my dad loved being a cop, he needed and wanted to bring in more money to support all of us. He still had his electrician's business and would do that during the day. Then he would work the midnight shift as a police officer. I remember coming home on leave from the Marines to visit and seeing how he was living and getting along. Dad would come home from work as an electrician, eat and take a short nap, put his uniform on, go and work all night, come home, eat, take a little nap, and then go do more electrical work.

In the process of trying to make money and put food on the table, working nonstop, he began taking ephedrine to stay awake.

★★★★★★★★★★★★★★★★★★★★★★★★★★★★★★★★★★★★★★

Working in politics this past decade, I've heard people say, "We don't pay our men and women in uniform enough." Oftentimes that's directed at our military. But meeting men and women serving as first responders, I've learned that same sentiment is very applicable to them as well. Unlike the military, where everyone of the same grade is paid the same, in law enforcement the pay is determined by the local municipality that employs you. The expectation of performance in this country is clear across the board. We expect our police officers to protect us at risk of their own life, to lawfully administer justice in situations not of their own creation and to do so flawlessly, but we don't compensate, reward, equip, or train them equally at all. I'm sure Justin's dad did his job with as much integrity and dedication as any member of the Los Angeles Police Department or NYPD, but in his small Indiana town, that same hard work and dedication didn't come with the same pay. I hear this often, especially in the fire service, where many work a swing shift, usually 24 hours on duty, then 48 hours off-duty, and have side businesses and jobs on their days off.

Unfortunately for Bobby Heflin and many other police officers in small towns across the country, using days off is not an option, so second jobs mean a lack of sleep and rest.

And like many people in our country, when sleep and rest give way to responsibility, we look for a way to stay awake and active. For some it's caffeine, for others it's street drugs like cocaine, and for many in the 1990s it was ephedrine. Over the past few decades the potency of ephedrine itself, along with its use as a main ingredient in illicit drugs, has resulted in federal and state laws regulating who can use it, when they can use it, and how often or how much they can use at a time.

PUSHING THE LIMITS

Somebody reported that my dad had exceeded the legal purchase limits of ephedrine and he was charged criminally. I don't know all the details, but he believed he was improperly charged. I've never looked into it, and I've intentionally avoided doing so. Like I said, my dad wasn't the best at politicking, and when he was criminally charged, I think he exchanged words with people in a way he probably shouldn't have. He always told people what he thought. Regardless, he ultimately lost his job. To add insult to injury, there was another issue where the department said that he had quit and wasn't fired. That meant he couldn't get unemployment.

With his relentless work schedule, all my dad was trying to do was stay awake and alert. I remember watching him take those pills. They came in a little blister packet. I never really thought about it. I knew people who took caffeine pills, and some people downed Red Bulls. It didn't seem any more harmful or beneficial than that. Except some people were using that same ephedrine to cook up meth, which you can't do with an energy drink.

On top of everything else, my dad was in pain. He suffered injuries on the job. He had to have multiple shoulder surgeries due to one incident. He pulled over someone on a motorcycle. They tried to flee, and he grabbed on to the handlebars and was dragged down the road. That was typical of him. He wasn't just a sit-around kind of guy. He had a reputation as someone to be respected, and he didn't put up with any nonsense. The thing that really hurt was losing his job the way he did, which got his name in the newspaper. As I've come to learn, having that kind of public attention focused on you as a law enforcement officer is a kind of kiss of death. He went from being a respected member of the community to being shamed and shunned by what seemed to be everybody. He struggled to find work. I think his pride haunted him for a long time after that. He had finally become what he had waited so long for. Once that was gone, the jobs he had after that just didn't seem to fulfill him. And the shame that he was put through without reason, I think that haunted him for the rest of his life.

I was in the Marine Corps at the time, so I saw all this happening but from a slight distance. I would listen to his stories and sympathize with him. But I didn't know what to do other than that. I was also building my own family, trying to make my way, and become my own man.

One of the biggest changes I saw in my dad was that he had become a drinker. My entire childhood, the only time I ever remember alcohol in our house was when he worked at the *Post Tribune*, a newspaper in Gary. We were painting our house, and we painted it together as a family. I was out helping. Some of his buddies from work came over and were helping paint, so he bought a case of beer, Miller Genuine Draft. We had a little outside refrigerator on the back porch. I remember seeing my dad drink a few beers with his buddies after painting. He would be giggly and silly. (Everybody has a memory of their dad just being silly, you know.) Then that was it. After the house was done being painted, I never saw him drink a drop of alcohol. I have no memories of my dad drinking other than that. He liked to drink Coke. That was always his drink, just some Coca-Cola over ice.

After Dad was dismissed from the police force, things changed. I was coming home from the Marines when I realized, *Oh, Dad drinks now, that's weird.* Later we would learn that he was drinking as a result of injuries he had sustained on the job. He would run out of pain medicine but couldn't afford the prescription refills. But he discovered that alcohol helped him, similar to what medications would help him for ailments, specifically his shoulders. Like I said, his shoulders were completely destroyed. He used to like golfing and such, but he couldn't hit the links anymore. He couldn't even throw things. My dad was only twenty years older than I was. For him to be hurt like that? I knew it bothered him. He couldn't work like he wanted to and he couldn't move the way he wanted to either.

The only upside, I guess, was that I learned to respect the dangers of alcohol from seeing what that substance did to my dad. I remember specifically one time when I came home from the Marine Corps. There I was, a young, proud sergeant of Marines who could drink beer with the best of them. And my dad got me so drunk I was throwing up off the back porch. But he was still going, and I was wondering, looking at him, *How can you do this, man? This doesn't make sense.* I don't think I had realized just how dangerous alcohol is, because I was never really exposed to it as a child. It was just something fun I did with my buddies. I never saw my parents drink. Then there I was, drinking with buddies in the Marines and having a good time. I thought I could drink with everybody, but I drank with a real alcoholic for the first time that night. That's when I realized there are different levels to alcohol consumption, and my dad's was much higher than mine. He indeed became an alcoholic and in a way lost everything as a result of the stuff.

You are held to a very high standard being a police officer. They told us in the academy that you no longer have a normal life when you become a cop. You live in a glass house. Everything you do is seen by everybody. People you don't know are going to know where you live, and they're going to point at your house as they drive by, and they're going to look at what you're doing. So you really do have to be a model citizen. And if you ruin it, it's ruined forever.

When I was a kid, my dad applied those high standards to us. When I was a teenager, my favorite sport was skateboarding. This was before Tony Hawk got superfamous so there were no skate parks when I started. Skateboarders had a kind of rebel reputation—antiauthority. I got shook down by a cop one time for skateboarding on a sidewalk that was marked NO SKATEBOARDING. I started running my mouth to the cop just being a punk, and so he shook me down pretty good and shoved me down on the car. He even got his dog out to scare me. I told my dad about it since I got a ticket. I had to go to court, and my dad had to take me there since I couldn't yet drive. He was pissed. I also decided to wear a T-shirt to court that said, I'M A SKATEBOARDER. ARREST ME. Dad about knocked my hair off my head for doing that.

★★★★★★★★★★★★★★★★★★★★★★★★★★★★★★★★★★★

Our parents have a profound impact on us. They teach us what to do, and sometimes through their flaws and mistakes, they teach us what not to do. While Justin and I both had alcoholic fathers, my dad started a lot earlier. He drank throughout my childhood. While differing in that respect, my dad, like Justin's, was strict and taught me to work hard, carry myself with respect and a respect for others, and do the right thing. So hearing Justin explain how his dad's mistakes and pain led to alcoholism was incredibly personal for me. My dad's ailments caused him to stop the grueling work of masonry, an occupation highly respected in the trades and blue-collar world, at a fairly early age. That left him with similar odd jobs to make ends meet. His pride took the hit as well. Although my dad drank almost every day after work, when work stopped, drinking became a full-time endeavor. The pressure of providing for a family with your body is a burden that many of us bear. And like Justin, I learned to pull from the good our dads taught us, but still be wary of their mistakes.

★★★★★★★★★★★★★★★★★★★★★★★★★★★★★★★★★★★★

ENTERING THE MILITARY

Tuesday, September 11, 2001, changed everything for me. Before that, I had no idea what I was going to do after high school. I knew I couldn't afford to go to college. I figured I'd get a construction job. I had no interest in the military. Again, I was a skateboarder with long hair. The military wasn't for me yet. But when we got attacked on 9/11? Everything changed. I also heard about Army soldier Jessica Lynch that year. That was big news. The internet was a very new thing, relatively speaking, but some of my buddies figured out how to watch videos online. I saw videos of the fighting and the bodies of dead American soldiers, which made me want to join and go fight. I remember watching the news and seeing the wounded being carried on stretchers and guys running and dodging bullets. I was one hundred percent as patriotic as it got. From that point on, I knew I was going to end up in the military.

Things almost didn't go as planned. Out of loyalty to a buddy, I enlisted in the Navy's delayed-entry program to be a medic, specifically a corpsman who works with the Marines. But I felt pressured to do it and instantly regretted it. I wanted to be a Marine like my dad was, but now that wouldn't be possible. I went home and I told my dad I had fucked up. He laughed and said, "Well, you're sworn in now, son. You're going to be a sailor."

As a last shot, I went to talk to a savvy Marine Corps recruiter. I walked into the office and explained my situation. The recruiter gestured to me to close the door. He then proceeded to write down something on a Post-it note. I read it, and he took it back, crumpled it up, and threw it in the trash.

Quite a few notes later, I had instructions about exactly what to say to get out of my Navy enlistment. The notes were used so the recruiter could say he never said a word to me about how to do it. And I had what I wanted all along—I was going to be a Marine.

The whole time I was signed up for the Navy, my stomach was in knots. When I figured out how to get into the Marines, I felt amazing. I could not wait to get to boot camp. As a bonus, my buddy got out of the Navy too and went into the Marine Corps. The Navy recruiters were pissed.

★★★★★★★★★★★★★★★★★★★★★★★★★★★★★★★

> I couldn't help but laugh when Justin said he didn't think much about the long-term when joining the Corps. He wanted to go to war but even with Iraq and Afghanistan popping off, not every Marine got the call to deploy. I get it. I had a similar experience myself. I was deployed to Iraq in my first enlistment, but not without a lot of maneuvering and volunteering. Justin didn't make it to war in his first enlistment, but fortunately for us both, the corps needed explosive ordnance disposal (EOD), or in plain English, bomb technicians. Like me, he reenlisted to do just that: deploy to Afghanistan and take apart bombs, or improvised explosive devices. But unlike me, Justin had a very personal motivation to combat these deadly things.
>
> Justin's sister, Maria, enlisted with the Army. While in Iraq,

the Humvee in which she and other members of her squad were patrolling was struck by an explosively formed penetrator (EFP). This specially shaped charge was designed to penetrate armor, and it worked as intended. Fortunately, Maria survived. But her fate only strengthened Justin's resolve to go to war himself. "That Humvee got a big, giant hole blown through it. That incident made us want to go to war even more. They tried to kill my little sister."

His brother, Ryan, also joined the Marine Corps. Justin said, "He was an infantry gunner for the Third Battalion, Seventh Marine Regiment, or 3/7, and he also got blown up on his first deployment. He was up in the turret. He spotted an IED and started to call it out, but before he could finish, he saw a giant fireball and his truck was over on its side. Now, it's kind of a joke in the family that my brother and my sister both got blown up, and I was an EOD guy and I didn't."

Fortunately, neither of his siblings was wounded in those attacks. And although Justin says he wasn't "blown up," he did two very eventful deployments in Afghanistan. His first was in 2009 and again as a team leader in 2011. I replaced his unit in 2010 when my unit took over the effort. Then Justin came back with his unit to replace us in 2011. That's how it worked for us in Marine Corps EOD. The west coast unit would go for a year, then the east coast unit would take over for a year. Most of our peers deployed every other year for more than a decade. That span between 2009 and 2011 was the deadliest of any war for Marine EOD techs. Justin had worked and deployed with Chris Eckard, a fellow EOD tech who was killed in early 2010 during Justin's first deployment. And like most Marine techs of that time, he remains grateful that his team leader, Cliff Farmer, was there to act as a mentor and friend.

★★★★★★★★★★★★★★★★★★★★★★★★★★★★★★

WORKING THE BOMB SQUAD

Cliff is at the top of my list of favorite human beings. Cliff gets zero credit for changing the way Marines went on patrol in Afghanistan by utilizing the "Holley Stick" as it was later named. It's essentially just a sickle he lashed together and took on patrol to save his back from bending over to check so many potential IEDs. Initially we just called it his sickle, but the name changed after Floyd Holley was killed while rendering safe an IED after relieving us from our deployment together. Floyd, who was commonly referred to as the sexiest man in Marine EOD, utilized the sickle while on patrol before being killed. Naming it after someone else shows the kind of guy Cliff is. He didn't need or want the credit.

At one point, Cliff and I were working around an abandoned school. The Taliban had taken it over, and we got sent out for an IED in the area. While we were sweeping around the school compound walls looking for one IED, we were being watched and had a remote-controlled IED detonated on us. I say "us," but more specifically detonated on Cliff. He couldn't have been more than ten to fifteen feet away from what we estimated to be approximately thirty pounds of homemade explosives. I thought for sure he was dead, but he walked away miraculously unharmed. Cliff was and still is a beast and a Jedi at the same time.

Another time, Cliff and I were doing our EOD thing, chasing down some wire in the hardened dirt of Afghanistan. Cliff and I often worked together on an IED, kneeling side by side, which was a huge no-no because our standard procedure was for a team leader to work directly on the IED while his teammate stayed back enough to have eyes on everything going on around him. This allowed for enough distance so that if the IED exploded the teammate would not get hurt. This time, we had a lot of area to cover, and Cliff said that I should

I've seen firsthand just how much weight a first responder carries even weeks after a tough call, and it has given me a whole new appreciation for and curiosity about the work they do. Having spent the past decade hearing their stories and seeing their faces around a campfire, I felt it was time to share their selflessness and sacrifice with the world.

Joey carried by Jeremy Judd on a hunting trip

Joey with singer Chase Rice (*left*) and Keith Dempsey (*right*)

From left to right: Joey with Clay Headrick, Keith Dempsey, and firefighter Justin Rishel

(All photographs courtesy of individual interviewees unless otherwise noted.)

★ ★ ★
CLAY HEADRICK

Growing up, Clay was my uncle's best friend, and they had a dirt-track race car they raced on weekends. Clay is a lifelong public servant who is willing to get uncomfortable to make a bad situation better—whether it's being the unsung hero on his best friend's race car team, working hard to take a legless buddy duck hunting, or sitting down with a brand-new firefighter in their moment of doubt to tell him or her that they really can do this.

Clay's official portrait

Training in rappelling

With his son, John

Teaching pressurized container fire control

KEITH DEMPSEY

Keith is more than a friend; he's family. Although most see him as quite reserved, when you get Keith talking about the fire service—the job and duty of protecting your hometown from disaster and tragedy—you see a true passion come out in him. A big part of his story was choosing to become a fireman over joining a proven family business that would have surely made him financially successful. To his surprise, his dad was all for it.

Keith with his wife, Tracie, and daughters, Harper (*left*) and Morgan (*right*)
(*Courtesy of Michael Cyra, Photographics*)

With his dad, David Dempsey

Keith and Gary Baggett on a fire scene

Keith in the Georgia Fire Academy with fellow firefighter, his uncle, Greg Garrison

One of the biggest catalysts for those who choose to be first responders is the mentorship and camaraderie they encounter early on. Keith and Clay learned from some of the best: Gary Baggett and Ronnie Dyer—veteran firefighters who poured their lives into others. Their tragic deaths echoed in their departments, underlining how strong the bonds are that first responders forge with each other.

Above: Gary Baggett teaching at the Georgia Smoke Diver (GASD) course. *Right:* Gary standing with an antique engine.

Left: Ronnie Dyer in uniform. *Above:* Ronnie graduating from the GASD course as a young man.

JEREMY JUDD

Jeremy is a proud veteran of the Maine Warden Service and claims one of the most decorated and storied careers in its long history. Simply put, a game warden's job is going into remote areas to check licenses for people they know are armed to kill. This sometimes leads to citing or arresting them. Sometimes it means hunting fugitives, and others it means helping families. When Jeremy was a kid, he was so inspired by meeting Warden Pat Dorian that he knew he had to make being a warden his life's work.

Above: Jeremy in diving gear talking to Mike Joy. *Right:* Jeremy training a K9.

With his grandfather Larry and father, Steve

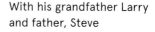

Reunited with retired Maine Game Warden Pat Dorian

Like many first responders, Jeremy worked more than just his assigned duty. He also served as a recovery diver for years, before the physical and emotional strain caused him to switch to K9 search and rescue. Together with a truly extraordinary dog, Tundra, he saved twenty-four lives. For Jeremy, Tunny was more than a dog or a friend; she was a way for him to move forward in his career, and go from reclaiming bodies to rescuing lives.

Tundra ("Tunny") and Jeremy

Jeremy and his family: wife, Melanie, and daughters, Annabelle (*left*) and Ava (*right*)

KATELYN AND JOHN ROBERT KOTFILA

Katelyn's brother, Deputy John Robert Kotfila, was a selfless hero willing to sacrifice his own life to save a stranger. For so many of us, losing a loved one to a specific thing might move us, with good reason, to avoid that thing. But Katelyn, honoring her brother's sacrifice, did the opposite. She joined the force herself.

Left: Katelyn and John Robert at her graduation, and (*below*) at an event

Katelyn receiving a flag honoring her brother

I can't imagine having a police officer for a dad, much less a detective. Not just because of the pressure to never get into trouble, but knowing he would find out anything I did—that would have certainly changed my high school experience. For Katelyn and John Robert, it meant learning a life of service early on.

Left: John Robert as a child, and (*above*) with his parents

TOMMY WEHRLE

I met Tommy on a bear hunt in northern Maine. His humility, coupled with his desire to be of service to others, really impressed me. Most of us go through life never having to consciously make a life-or-death decision, or having the choice to take a life to save another life. But that's what SWAT operators like Tommy have to do, seemingly every day.

Left: Tommy with his wife, Nicole, and children, Max Bosley-Smith (*right*) and Hunter Wehrle (*left*)

On a counterterrorism course in New Mexico with his team

With teammate Anthony DeMarnio (*right*) on the latter's last mission before leaving SWAT

Friendships take many forms. For me, hunting trips with these men were the perfect setting to learn how similar the experiences of first responders and veterans can be. Often, those experiences literally overlap. My friend Justin Heflin is one of those heroes who hung up camouflage utilities for a trooper hat and badge.

From left to right: Joey with friend Greg Wrubluski, Tommy Wehrle, Keith Dempsey, and Dave Hentosh

Helicopter rappelling training for Tommy's team

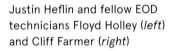

Justin Heflin and fellow EOD technicians Floyd Holley (*left*) and Cliff Farmer (*right*)

Justin's sister, Maria, standing next to her squad's Humvee after it was struck by an explosively formed penetrator

JUSTIN AND BOBBY HEFLIN

Father-son relationships can be complicated. For Justin, growing up with a father, Bobby, who was a Marine and a cop meant that despite a youthful rebellious period, Justin would be inspired to become a Marine himself. But the mistakes his dad made also served as a stern warning of the cost paid by those who risk their lives and health to keep society safe.

Above and right: Bobby Heflin with son Justin on a base in North Carolina

Justin skateboarding as a kid

With his parents before he left for Okinawa for two years

Justin's dad, Bobby, did leave one legacy that had a profound effect on Justin: he taught him the toughness and ambition he'd need to make it as a state trooper. Justin literally followed in his dad's footsteps, first into the Marines and then into law enforcement.

Justin introducing his son to his dad

Justin with a shirt he won in a USMC pull-up challenge at the Indiana State Fair

Justin with friends and fellow EOD technicians Johnny Morris (*left*) and Dustin Johns (*right*) at the Indianapolis 500

VINCENT VARGAS

Army veteran, actor, entrepreneur, and journalist Vince Vargas has lived all sides of the immigration issue, not just as a Mexican American himself, but as someone who has worked as a Border Patrol agent. As a medic, he worked to protect every life he encountered.

Vince as a ranger school honor graduate with his uncle Mike

In the army

Serving as a Border Patrol medic

With his family: wife, Christie, and kids (*from left to right*), Belle, Jarek, Holden, Ryker, Hunter, Star, and Taylor

★ ★ ★

STEVE HENNIGAN

From responding to bomb threats in Los Angeles to resolving domestic violence situations, Steve Hennigan's career covered every sort of circumstance. For all of that, he responded to these perils with a shocking amount of both courage and empathy. Respecting others was his motto, and he upheld it.

Top left: Steve with his mom at his graduation. *Left:* Steve with a patrol car. *Above:* Steve holding a tiger cub in his motor officer days.

An official photo near Steve's retirement

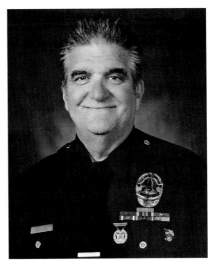

Sometimes the first responder life can get a little Hollywood. Literally, in Steve's case, as he would sometimes run into celebrities on the job in Los Angeles. Vince has made a career in TV and movies. His first film, Range 15, featured several notable vets (including myself) as zombies. Sheriff Mark Lamb may have utilized his towering stature and charming personality to gain a bit of a celebrity following on social media, but the work he does is all too real, and his efforts to reach the press with the truth about policing America's cities are deeply meaningful and compelling.

Steve with actor R. Lee Ermey at a celebrity golf event

Above: Joey and Vince on the set of the film *Range 15*, and (*left*) at the gym

Mark Lamb and his wife, Janel, visit Joey on the set of *Fox & Friends*

★ ★ ★

MARK LAMB

Mark Lamb's passion and demeanor are everything you'd want to see from a leader in law enforcement. And it doesn't hurt that he stands well above six feet tall and effortlessly sports a cowboy hat. He gives off a humble confidence that makes the word <u>lawman</u> come to mind. But Mark is much more than just an image. He's a dedicated public servant.

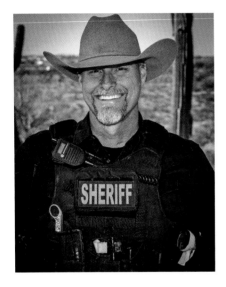

go one way and he'd go another. Long story short, he encountered an IED. "Hold on, I got something," I heard him say. So I started walking toward him. I asked him to tell me what he was seeing as I approached him. But he told me to stop, which was unusual for me and him, and I did. From a distance, he was describing what he was looking at. Long story short, he said it looked weird, and he wasn't sure what he was dealing with just yet.

Turned out, that IED was booby-trapped like the one that had killed our friend Chris just three days prior. This one has always haunted me. Cliff figured that one out and disarmed it. He was way more experienced than I, and a much better EOD tech. Had he not told me to go the other way on that call, I would have been the one to come up on that thing. They probably would have gotten me, or at a minimum, rocked me pretty good. It definitely made my hair stand up. We had quite a few experiences of being lucky during that deployment.

I left the Marines in 2012 and went back home, where I immediately applied to the local police department. I actually finished number one on their list after taking the PT test, the written test, and doing the interview. Then I had to take a polygraph test. A little while after that, I got a rejection letter. I didn't get any explanation about why I was rejected, just that I was rejected. I figured it had to be something with the polygraph, but I couldn't figure out what it could be. I was completely honest. I had a clean record. I even admitted to them that I had smoked marijuana in high school. This was nine, ten years ago, and maybe that disqualified you then. I wasn't happy with how things turned out, and I ended up becoming a union electrician and did that for a couple of years. I thought I might have made a mistake in leaving the Marines and so began the process of going back in. I even talked to a recruiter, took some tests, and began preparing for it physically.

When a gunnery sergeant pulled rank on me during a conversation, it brought back memories of why I got out in the first place. That conversation ended my quest to go back to the military. But I also wasn't completely happy with being an electrician. The money was good and I had a lot of fun with the other guys, but I couldn't imagine myself at sixty being on a scissor lift running three-quarter-inch conduit across a grocery store under construction. I just couldn't see doing that for the rest of my life. I also couldn't see being a cop, because I'd already been rejected. So I thought that was gone. Driving home from work one day, I saw a state trooper parked at a local college.

As chance would have it, like me, that trooper was ex-military. He saw the base sticker on my pickup truck and started asking me about it. He told me that the state police had their own bomb squad and that I should apply for a job. I was reluctant, but I eventually did apply and go on a ride-along.

I was hooked. I was blown away by what the life of a trooper was like. We were chasing guys down the highway at 120 miles per hour, making stops, and taking guys to jail for drugs. And they got paid to do that? And I could be on a bomb squad? *Again?* I knew there was more to it than that, but comparing that adrenaline rush to the tasks I completed as an electrician? There was no competition.

I applied to the Indiana State Police, which was a very difficult agency to join. There were guys in my academy who applied five times before they got hired. I was very lucky and got picked up on my first try. No college, but a military background, and I got picked up and went to the academy, where I graduated as a trooper in 2014.

Besides the adrenaline rush of the job, though, what appealed to me about the work is the camaraderie. They looked like they had a lot of fun. Seemed to be everybody was sitting around breaking bread and kind of cutting it up. It felt like what I did in the Marines as an EOD tech. There were days when Cliff and I would be in the chute first. We would come back to an empty squad tent and wait. Another nine-line call would come in and we would jump up, grab our gear, and go. We were a quick reaction force, and we were there to protect our guys. I was trained and capable of shooting, and while I did fire my weapon at times, the main job was to preserve lives. It's the same thing as a trooper. And every day was different.

On any call for service, whether it's somebody stranded or lost, or a rolling domestic going down the road, or a stolen car, we're an emergency response. When it was time to go, it's *go right now* and you're off like a shot. You know it's action time. And of course, because my dad had been a cop, it felt like it was in my blood.

> Not unlike when we got an IED call in Afghanistan, I'm sure the adrenaline takes over when troopers get a call to respond. But with time and experience comes an understanding that this job isn't all about simply taking out bad guys. It's not always that simple, or that fun, or even that black-and-white.

"NOT WHAT YOU KNOW, BUT WHAT YOU CAN PROVE"

I had a young lady once who I was *sure* was being trafficked. I pulled a guy over driving a semitruck and pulling a trailer, a pretty routine stop. At least at first. I climbed up on the step of the cab talking to him. I noticed the curtain was pulled on the sleeper part of the truck. I asked him if anyone was back there. He was being evasive and awkward. Eventually he did pull the curtain back. There was a young lady in there. She wasn't a child. She was an adult. But still a young lady. Early twenties, probably. And he's obviously older than she is; he's a grown, middle-aged man with this pretty young girl in the back. So right away, that's a little strange, and she was a little dirty-looking. It looked like she'd been living in the back of the truck. She had no shoes, dirty feet, and was wearing a dirty sundress. He said that he used to be her stepdad. He had divorced her mom, and now the two of them were friends.

When I asked for identification, the guy told me that he had hers. Alarm bells were going off in my head. How she looked, no shoes, not carrying her own ID. And the guy was trying to laugh this off and minimize my questions to her about why she didn't carry her own ID. I remember thinking that she had to have a purse. All women carry a purse, right?

Everything about it was awkward. So I asked him to get out of the truck and separated them. I requested that a female officer come talk to her. I tried talking to her myself one-on-one, and I realized that maybe she didn't want to talk to me because I'm a man.

The female trooper came out and talked to her and said everything she could to her to try to make her feel comfortable. But still the young woman said, "No, guys, there's nothing weird going on. We're friends."

I got back in the truck and talked to her one-on-one. I told her she just had to tell me she needed help, but she was adamant. I kept thinking back to a training I had with a lady who was trafficked. She was trafficked to the point that she told us that her body was rotting away, that pieces of her were rotting off. She lived in a cage and was traded for a car at one point. She told us that while she was being trafficked, she had met police officers in Vegas, and the police officers in Vegas looked at her and asked her questions, and tried telling her, "Yo, like, hey, what's going on here?" And she said she refused to answer. She refused to allow herself to be helped.

That's all I could think while I was talking to this girl: "You won't let me help you. I'm here for you. Please, let me help you."

In my heart I think there was something afoot. But just because I know it in my heart that doesn't mean I can haul someone away to jail. There was nothing I could do to help the young woman. I gave the man a ticket and they drove away.

Maybe I'm wrong, and everything was on the level. But why on earth would a former stepdaughter be riding around the country with this guy in the back of his truck, pulling trailers for money, with no shoes on? Obviously she hadn't bathed in days, with a dirty sundress, and he had her ID. I don't know anyone who would consider that normal.

I had no choice but to send the pair on their way. They were out of sight but not out of mind. Seeing things you can't unsee is in the job description when you work in law enforcement.

When you go to bed after seeing some pretty messed-up stuff, you see those scenes again after you close your eyes. Your heart will still be breaking for some of the people you interact with. It could be as simple as being in somebody's home and witnessing how some kids have to live. They have parents who obviously don't care about them, which hurts to see. You always remember the kids who are affected by families who are doing wrong things. I've had to make death notifications to families. Maybe it was a suicide, maybe an accident. You make that death notification and you're hugging the mom of the victim, the deceased. You always remember the mom. And there's times when you're hugging mom, and you look over and see her other kids. They're staring at you and looking at you like they hate you. They're just mad that you're delivering bad news, but still, you remember that.

★★★★★★★★★★★★★★★★★★★★★★★★★★★★★★★★★

The psychological toll of being the one charged with protecting the innocent and oppressed, but not always being able to do so, is heavy. After having my daughter, it got much harder to report on stories where children were abused or hurt. Those

> stories would stay with me for days. The fact that Justin doesn't just read them on a teleprompter or interview their loved ones from afar, but instead has to face the crimes and the families to tell them what happened, shows how thankless this job can feel.

MEETING PEOPLE IN THEIR HARDEST MOMENTS

Sometimes being on-scene when a person dies presents another challenge. I had an overturned truck one time. I was first on-scene. The driver was pinned against the overpass, upside down. I got there and crawled my way over to him. I was trying to figure out how to get this guy out, and I knew there was no way I could do it.

He's in the cab and we're talking, and he's crying. It's me and him, and I'm trying to keep him conscious and calm. I ask him about his family: "Hey, c'mon, man, you got a family, right? Tell me about her." Turns out he did have a wife and a family. I keep asking questions, trying to reach into the cab. All I can do is get one hand in there. He's upside down and all I can do is touch his belly and talk to him. The truck is mangled and I'm in this physically awkward position. I'm talking to him one minute and then the next minute he's dead. I can still hear his voice.

Another time, I arrive on-scene. A man is driving along the interstate with his parents. His parents are in the back seat. His wife is in the passenger seat in front. They get a flat tire and pull over onto the shoulder. While he's still inside the car, another vehicle crashes into his. The car with the flat bursts into flames. The driver manages to get out. No one else does. I get there and the car's burning, and the guy is understandably distraught. He wants to go to the car. I have to talk him back and tell him that he doesn't want to do that. He doesn't want to see that. The car is burned up, and I'm there on the shoulder trying to keep him distracted from this catastrophic scene. He doesn't care what I have to say. He wants to get to his family. You're trying to do the best you can to keep him from having to see that. But I have to see this whole thing going on, in the moment and later, and for a long time after. It's my job to have to walk up to that car and witness all that. Later you'll hear something, see something, smell something, and it will take you right back to those moments. A lot of police work, I think, compares to experiences of the military. Seeing carnage doesn't affect you differently if it's on a deployment than it does on your regular duty shift.

Sometimes those events hit closer to home than others. I had just moved. I had a great next-door neighbor at the old place. Heck of a great neighbor. His wife had died a few months before this. He was planning on selling his place. He had a dumpster out front and was cleaning things up for the sale. I get the call about a death investigation. I knew the address well since it is right next to my old house. In the garage, sitting on the lawn chair, is this guy, a great guy, dead. And I'm there in an official capacity helping with a death investigation. I'm back in my old neighborhood talking to people I knew. And I'm there being professional, doing my job, and there's this guy who used to give my kids Christmas presents. My wife had baked Christmas cookies for him. We used to stand along the fence line and chat. He used to yell at my dog and tell him to shut up.

Then I talked to his kids. Later they said they were glad that I was there with him. Sometimes you later go by a house like that, or the area of some other incident, and of course you think about that moment. It's not like it changes my perspective of the area. I live in the town where I spent most of my time growing up, so I have memories of running around, playing, climbing the trees. But there are those other incidents where people had a really bad day and I was a part of that. It doesn't make me want to move away or anything, but it just crosses your mind. It simply becomes part of your memory.

You can't completely prepare somebody for seeing and living certain moments. But because I had been in the military, I felt I was better prepared than some others. The first time I saw a dead body wasn't when I was a police officer. I'd seen one before, and I knew how to detach, be involved, but detached. I knew that as a trooper, I was getting in bed with a lot of ugly. So there's the thrill part of it, the adrenaline-rush part of it—catching the bad guy, saving the girl who needs help. But there's also the bad part. And if you come into police work not knowing that you're going to deal with the bad, you're naïve. I believe all cops know, coming into the profession, that they are going to see things that won't make for great dinner-table conversation. You can tell some stories, but you can't share all of them. Some of that stuff you can share with other cops, and that's it.

NEVER OFF-DUTY

On one occasion, my wife and kids got a front-row seat to some police work. It was a Saturday, and I just got done cutting the grass. We had some errands to run, so I piled in the car with the family, and my wife was driving.

I'm in a pair of cutoff cargo pants and an old Marine Corps T-shirt. I'm a mess, covered in grass clippings. My wife takes the car into a drive-through car wash. We're just barely creeping out of the finishing touches of the car wash when I see a car pull up, obviously speeding and going the wrong way into the car-wash property.

Two seconds later, another car pulls in. I tell my wife to be careful while pulling out. I watch for a second as a young black male carrying a tire iron approaches another young black male who has a full-term pregnant woman in front of him, and he's got his arm around her throat in a chokehold-style situation. I tell my wife to call 911 and I'm out the door, using one hand to draw my gun and the other to pull out my badge. Running with my weapon drawn, I start yelling at them, "State police" and "Everybody down on the ground," as loud as I could. There are about a half-dozen people congregated there. I'm still screaming at everybody to get down, my wife's yelling at me from the car, and out of the corner of my eye I see my son trying to open the door to get out there to be with me.

Then I hear the sound of a car horn blaring. The guy behind my wife in the car wash is upset that she's not pulling forward. He's confronting her. Then another group of people are walking through the car wash toward me, telling me that that's their family I have there on the ground. I yell at them to stop approaching me, again identifying myself as state police. During all this, I hear the old white guy whose truck was behind my wife's car yelling at her to move her car out of his way. She's still got the phone to her ear, trying to tell the dispatcher everything that's going on. She wants to make it clear to the cops that show up and see me with my weapon drawn and not in uniform that I'm not the bad guy.

Then the upset old white guy decides to take his complaint to me and not my wife. So I've got all these people on the ground at gun point, and he's telling me to tell my wife to move her car so he can be on his way. I looked at him and said something along the lines of, "Sir, I'm not dealing with you right now. I'm state police. I need help. I need more cops. Call 911 if you want to."

Then, from the ground, I hear one of the young black guys say to the old white guy, "Are you fucking serious right now? You see what this man is dealing with, and you worried about getting your car out?"

I'm keeping this black guy on the ground at gunpoint, and he's defending my actions against this white guy who is doing nothing at all but being a hindrance.

Eventually the local police arrived on-scene. Just before they arrived, I had another exchange with the young guy on the ground who had intervened on my behalf. He was lying chest-down on the ground in the Superman position and wanted to get a cigarette out of his pocket. I told him he could, but that if he pulled out anything but a cigarette, it wasn't going to be good for him. He got a cigarette out and a lighter and lit up. He took a puff and said to me, "Man, this shit is crazy." We both kind of laughed about the situation.

After the local officers took charge of the scene, I went to speak to the impatient car-wash man. I had a business card in my wallet and gave it to him. He asked me, "Are you even old enough to be a state trooper?" I ignored that. Instead I told him that he and I had nothing productive that we could say to one another. I told him that he could have helped me, but instead he made things worse. Now he had my information, and if he chose to, he was welcome to speak to whomever he wanted to and lodge his complaint. I got in the car with my family and drove away.

That was really crazy and chaotic, but more frequent are the times when you're in the grocery store with your wife and kids and you see a bunch of other knuckleheaded kids in an aisle, and you know they're up to no good. You have to decide: *Is this worth paying attention to? Or do I just skip that aisle?*

Your cop radar is always on. You develop a sixth sense, a situational awareness that is hard to turn off. And living in the community where you work, you have to be aware of every move you make. Can you go out and have a few beers with friends? Is it worth it?

It's strange having to be so conscious of all your actions all the time. Having to consider the consequences of every move you make, every word you utter. And the repercussions don't just affect you but

THE CATCH-22 OF POLICE WORK

your family. I've had threats made against my family. I haven't worn a wedding ring since maybe 2014, so nobody will know that I have a family. Once I had someone I dealt with tell me that I looked to be about his age and that he could find me and my family. There has been an increase in threats against the police generally in the last few years, and I park my police car outside of my house, as most officers do. So there have definitely been times when I've slept with one eye open.

★★★★★★★★★★★★★★★★★★★★★★★★★★★★★★★★

It's a shame that Justin and his family have to endure those threats. The vast majority of law enforcement officers go into the field truly wanting to serve and to protect. In the aftermath of George Floyd's death, antipolice sentiment was high. All cops became the target of the ire of some, and those threats that Justin faced had nothing to do with any actions he'd taken. They were the result of him making a choice to wear a uniform and a badge and face the possibility that he might lose his life or well-being in service of others.

Justin wanted to go to war to protect his country and knew that if it came down to it, he would be called on to take another person's life. He would answer that call without hesitation. In doing police work, his attitude toward the use of lethal force again reflects how nearly everyone in law enforcement feels.

This country has endured a certain antipolice movement in past years. The reporting on events like the deaths of Michael Brown in Ferguson, Missouri, and George Floyd in Minneapolis triggered what appeared to be a powder keg of violent youth and others setting fire to their own towns, as if such actions would stifle what they called "police violence." But what we don't report on is the rippling effect it has had on police officers around the country who do their job with honor and integrity every day. From selfishly ignorant people like the man who approached Justin in the car-wash incident, to the pathetic cowards who would threaten him and his family because he arrested them when they broke the law, the seemingly growing sentiment that "all cops are bastards," as protesters commonly shout, becomes a stress and danger they should not have to endure. Like most any police officer you'll encounter, Justin has gone to great lengths to put his own life on the line and protect the lives of those he serves. Even when it comes to protecting the suspects he's arresting and using his discernment on whether to use justifiable deadly force, it is his care and concern for all life, and not political pressure, that guides Justin's compass.

★★

LIFE UNDER A MICROSCOPE

I never want to hurt anybody. I don't want to fight anybody. I don't want to kill anybody. More importantly, maybe, I don't want to get hurt, and I don't want to get killed. In 2022, law enforcement was still under a microscope. There was a pursuit taking place about fifty miles from where I was working. A trooper was in pursuit of someone and had shots fired at him. The car the shooter was in was headed in my direction, so I got on the road and started going toward him. The other trooper stopped chasing the guy. I was thinking, *If you're shooting at a trooper, I'm coming after you. I'm going to find you and stop you.*

The guy ended up crashing his car. He took off on foot and disappeared into a wooded area near the interstate. I was heading out there, lights and sirens, getting there as quickly as I could. I met up with a classmate of mine on-scene. He was another trooper, a canine handler,

THE CATCH-22 OF POLICE WORK

and the two of us came up with a plan real quick. We knew the direction the guy had headed, and after putting on extra body armor and getting our rifles, we headed out with the dog. We started scouring the area, kicking every bush and talking to everybody that we saw. We were looking for a black male dressed in green. Nobody had seen anyone fitting that description.

We made our way back to the interstate. A helicopter was circling overhead. We learned that the guy who had fled was wanted on murder charges and was known to be violent. Well, he had demonstrated that by taking shots at a trooper. We went out again after a short helicopter ride, continuing our manhunt. We came up on a darker-skinned male, a Hispanic guy sitting naked on the ground under a bridge. He was sticking a needle in his hand. As we approached, we issued loud, clear verbal commands for him to show us his hands. We were along a river, and the guy jumped up and ran into the water. We saw that he was armed with a pistol in one hand. I had him dead to rights from my position as he was in the river.

You think of a million possibilities in that moment. I set out to find a black male in a green shirt. I found a naked Hispanic male. He has a gun, but is he pointing it at anybody? I was up there watching him tread water, and he's shouting at me to shoot him. He's telling me to shoot him in the head. And I was shouting back at him telling him that I was going to shoot him unless he drops the gun. I can clearly see the gun isn't pointed at anyone. There's nobody around and we're in the middle of nowhere. My weapon is off safe position. I yell to the other trooper that the guy has a weapon. I'm watching the dude; my finger is on the trigger. There's open terrain between us, and I know I've got a clear shot. I'm also thinking that I have my body camera on, and if I do fire, this has to be as legitimate a use of lethal force as can be. I tell myself, *He's not pointing the gun at anybody, but as soon as he moves his hand with that gun, this guy is dead.*

Well, the guy stops treading water and starts swimming downriver. We pursue him and I'm wondering, *Is this the same guy who shot at the trooper earlier? Is this the murderer who took shots at one of our own? Or did we just scare a homeless guy living under a bridge?* I also now hear that other guys are joining the two of us in this pursuit. I climb over a barbed-wire fence that runs along the river. My buddy has to take a different angle of pursuit because of the dog and the fence. The helicopter is now circling above us. There are other cops up and down the river from our position. The guy is in the river, not swimming anymore, but settled in one spot. This whole scene is chaotic and noisy, and we're in a kind of standoff, and I wonder how long this can go on. I keep thinking that our suspect needs to get out of the water. He's got to be freezing. So, I asked if the helicopter could be moved from the area. I handed off my rifle to someone else and made my way down to the water's edge. I just wanted to talk to the guy. Now it was quiet and I knew he could hear me. Maybe his adrenaline level had dropped and he could make good choices.

I said to him, "Hey, man, stop this. What are you doing? Dude, it's just me. Just me. I didn't shoot you earlier. I don't want to shoot you now. Get out of the water, man. Get yourself warm." Finally he did surrender.

It turned out he was the one who had fired on the trooper earlier in the day. He was the man who was wanted for murder. But if I'm going to take somebody's life, I know that I'm going to have to answer for every action. I'm going to be investigated for a homicide. Even before bodycam eras, I held myself accountable. I've always tried to do my job so that my kids and my wife would be proud of me.

That can be tough, given an increase in people who purposely bait police officers into confrontations. There is no shortage of people that want to put stuff up on YouTube showing how bad cops are and how poorly they handle some encounters with the public. The

I-don't-have-to-give-you-my-ID-if-I'm-not-under-suspicion-of-a-crime-or-if-I'm-not-being-formally-detained routine never ceases to amaze me. I always have to laugh to myself when that starts up, because I think, *Dude, I promise you I'm not messing with you if I don't have to. I'm not losing my job for a ticket or an arrest.* I have never witnessed another officer risk his for it either.

Police in America are the best police in the entire world. Maybe my opinion is biased, but based on what I've seen in other countries during my travels, I'll put American law enforcement as the most just and righteous in the world. But you'll still hear "I hope your mother gets cancer!" or "I hope your wife cheats on you!" or the pleasant "I hope your kids get sick!" The unnecessary and uncalled for disrespect just blows my mind sometimes.

I started at the academy on my son's fifth birthday and had to miss the celebration. And when I was there, they talked about imagining that in every interaction you had, you had your kid on your shoulders, so conduct yourself accordingly. That really resonated with me then and it still does. It's always in the back of my head.

I don't want to disappoint my family or disrespect my uniform. I always do a check: Would my family approve of my actions right now if they could see me? Would they be proud of me or would I disappoint them?

THE COST OF SERVICE

My dad was, to me, larger-than-life. He was strict. I loved him. I respected him. I wanted to be just like him. I followed in his footsteps. But in the end, alcohol got hold of him and never let go.

One night I was on a midnight patrol. I had a scanner that allowed me to pick up radio calls from the local police. I heard about a pedestrian-and-vehicle crash not too far from me. I moved out to assist them. On the way, I spotted a speeder and did my job. Typical trooper, always looking to run some traffic. I pulled them over and issued a ticket. Awhile later, I met up with the other officers and heard about that call I didn't go to.

So, they painted the picture for me. A lot of gallows humor about it, like usual. You talk about it that way so you get it off your chest. Because you're gonna see it at night. When you close your eyes, you're gonna see that forever. They said it was pretty bad. The guy on a bike got hit by a car and didn't make it.

I didn't think much about it. I had an arrest later that shift and had to transfer a prisoner to jail two counties away. My sister called me and asked if we could meet up. I wanted to know what was going on, even though my gut was already telling me it was about my dad. I was putting the pieces together and figured that crash I didn't get on-scene to involved my dad. I knew my dad was dead. I was right. He'd been hit by a car while riding his bike. He was just fifty years old. His alcohol levels were three and a half times the legal limit.

I didn't really know what to do. I had this guy in my car who I had to deal with. I wanted to get back home and be with family and attend to whatever needed to be attended to with my father. *My dad is dead.* All these thoughts were running through my head. I had to finish the job, but I called Command and explained the situation. It all got handled. The state police took great care of me that night. Everybody told me to get home and be with my family. I met with my brother and sister and, while still in uniform, did the death notification to our mother and extended family.

There was one other person I needed to see, and that was Dad. I went to the hospital with my brother, Ryan, and they arranged that for me. We spent a little bit of time with him, then went home to our families. Later I heard from some of the guys at the police department that when they got on-scene with my dad, there was a teenage kid, a bystander, who was sitting there with him, talking to him and trying to help. Kind of like what I'd done with that guy in the flipped semitruck.

I got the name of the kid, I don't remember his name now, but I spoke with him and thanked him for being there while my dad drew his last breaths. I was glad that he wasn't alone. I met up with the teenager and his mom, and to thank him for what he'd done, I brought him some of my dad's Marine Corps stuff. I gave him one of my dad's eagle globe and anchors and a set of his sergeant stripes, some of my dad's prized possessions. I also gave the kid my phone number and told him that if he ever needed anything, he should call. I owed him one. He was obviously shaken up by what he'd seen, which was absolutely a heavy scenario, especially for such a young man, and he started crying. I hugged him and thanked him again for being there with my dad. I felt bad for him having to be a part of that, but I was also grateful. I've often wished I would have gone to that call that night to be there with my dad. I don't know. Maybe it was better that I wasn't. But I'm happy that someone was with him.

My dad was the man who taught me so much and steered my life toward what it is today. My family eats dinner together now, and I'm the one holding formations without them knowing it. We clean our house together, and they don't know it's a secret Marine Corps field day. And now my kids have their own memories of watching *Cops* together as a family. It's difficult to say whether I would have become a police officer without my father's example, but I know I'm forever grateful for his influence leading me to the Marine Corps and the Indiana State Police.

This is the side of "serving your community" people rarely see or hear. Perhaps that's just because they don't want to. Maybe it's because for many of us, saying goodbye to a loved one means seeing them "made comfortable" in a hospital bed, or perfectly dressed and groomed in a funeral parlor. But for our first responders, it's often gruesome and traumatic seeing someone die, and for those who serve the community they live in, every call they respond to comes with the fear that it is someone they know and love who is involved. To see Justin's life come full circle, from watching his dad work hard to be a police officer and then becoming one himself to nearly responding to his dad's death, puts an exclamation point of reality firmly at the end of this chapter. You may have heard that saying, "Not all heroes wear capes"? Usually it's a social media comment referring to someone doing something that might get themselves in trouble but is objectively the right thing to do. Hearing these stories from Justin and hearing him explain how it feels to deal with the messes we as human beings make, I don't know how he gets up every morning, puts his uniform on, and continues to go to work. But he does. Perhaps not all heroes wear capes, but a good many of them wear badges.

PART 3

DUTY OF DISCERNMENT

★ ★ ★

Reading people is really the hard part of the job. That's the stressful part, trying to figure out these people and what they're going to do, or what they're capable of doing, because you just don't know, and there's no book that you can read or teacher that can tell you.

—*Clay Headrick*

It's hard having to be so conscious of all your actions all the time, having to consider the consequences of every move you make, every word you utter. And the repercussions don't just affect you. I've had threats made against my family. I haven't worn a wedding ring since maybe 2014, so nobody will know that I have a family.

—*Justin Heflin*

How many people really understand Border Patrol? Very few people understand that every situation you come up on, you have to make a decision about which hat you're going to have to wear. Are you saving a life? Taking a life? Consoling a family member who just lost a loved one? Are you providing medical interventions? Or are you on a humanitarian mission? So, it's very complex.

—*Vincent Vargas*

The Watchdog

★ ★ ★

VINCENT VARGAS

★ Border Patrol ★

"In the military, I was an instrument: I was trained to kill high-value targets. In BORSTAR, my job was the opposite: saving people."

When I tell people I grew up in a small town in northwest Georgia, they paint a picture in their minds; maybe they know about our beautiful mountains, or perhaps they picture the Atlanta skyline. But rarely does the picture they conjure match the description I give next. I tell them Dalton, Georgia, is a beautiful textile town in the foothills of the Appalachian Mountains. Not just a textile town, but in fact the carpet and flooring capital of the world.

As the conversation continues, I get to the real stunner, which is that with a demand for labor even beyond that of our agricultural neighbors just south of us, over the 1990s and early 2000s Dalton, my deeply southern hometown, grew to be upwards of 60 percent Hispanic. So when I share opinions about the problems at the border, I have faces and names that go along with them. Take these two guys I'll call Diego and Daniel, who worked manual labor for my dad on his masonry crew. Diego's dad came to Dalton from Mexico illegally in the 1980s or '90s. He met Diego's mom here in Georgia. Diego was born a US citizen and attended our local community college, where he earned an associate's degree. Daniel was Diego's half brother and was born and grew up in Mexico. It is my basic understanding that their dad had a family in each country.

Diego spoke perfect English, but Daniel spoke only Spanish. So, to help his illegal brother work, Diego sacrificed the career opportunities available to him as a college-educated US citizen to work alongside and translate for his brother. My dad respected Daniel for this. He also paid him a dollar an hour more than his white peers because he had to both do his job and help instruct his brother.

As heartwarming as that is, I have several more faces that

go along with the massive amount of illegal immigration that descended on my little boom town. When I was in middle school, my uncle Jeff, who lived next door, came over late one night and asked me to take him on my four-wheeler through the trails behind my house to go look for a man who had hit-and-run at the intersection near the end of those trails and fled on foot. My uncle's friend was a local sheriff's deputy and was responding to the accident. Uncle Jeff knew about the trails and feared the man might use them to get away or worse, follow them to our family property.

We followed the trails down, me driving and Jeff armed with a .22-caliber squirrel rifle. The trails ended in a vacant lot between a convenience store and a daycare. Following common sense and intuition, we slowly made our way over to the back of the daycare and there the driver was, crouched down, hiding. The sheriffs eventually arrested him. He was illegal and had been drunk driving with no license, insurance, or registration.

A few years later my same uncle Jeff was getting in his truck at that same convenience store when he heard a loud crash. He looked up and saw that a vehicle had run the (only) stop sign and T-boned another car. He quickly ran up to the cars, which were still entangled. The hit car was our community church van and the lady who played the piano was driving. She was shaken up but seemed okay. He turned to the other vehicle, a Chevy Tahoe, that had hit her. The man driving was drunk and speaking only Spanish.

The SUV was still running and the man seemed to be trying to untangle the cars and flee by backing up the vehicle, then speeding forward, hitting the church van over and over again. Jeff told him to stop, but the man hit the gas pedal again.

Jeff reached in to grab the keys but was only able to turn the Tahoe off. He told the guy to get out of the car, and when the man tried once again to flee the scene, Jeff "assisted" him in exiting the vehicle and "gently" laid him flat on the pavement until the sheriff's deputies arrived on-scene.

Although it may seem my two best examples paint this specific intersection and my uncle as uniquely poised to intercept the rare car crash involving illegals, it's more evidence of just how frequently such things happen in our town. In fact, most people I know here have similar stories.

These examples represent the true duality of illegal immigration and the impact it has on communities all over the country. For every Diego and Daniel simply trying to earn a living, there are countless others who are completely disenfranchised from American society, living in the shadows, committing countless crimes and endangering lives with reckless behavior.

Much like most of the Texans you'll meet, we love many of our Hispanic neighbors we grew up with, most of whom are first-generation Mexican Americans. And if you want to know their politics, it's one of the strongest Republican districts in the country. They are conservative and vote that way. The issue of immigration doesn't just resonate with me; it impacts my daily life and tugs on my heart. I care about the people affected by it because they're my family. I care about the people whose parents came here illegally because they are my lifelong friends.

One man I've come to call a friend knows this issue in ways I'll never personally understand. Army veteran, actor, and entrepreneur, Vince Vargas has lived all sides of this issue. Not just as a Hispanic American himself, but as someone who has worked as a Border Patrol agent and now journalist.

CHASING THE AMERICAN DREAM

I grew up in Los Angeles County in an area called the San Fernando Valley. I was born in 1981, and in my younger years, the gang presence was very prevalent in the Chicano culture. What I didn't realize then, but did as I got older, was the struggles I had with understanding my culture and my identity and how much people would lean on that in how they perceived me and how I perceived myself. My mother is first-generation American. Her mother, my grandmother, came across the border illegally. She used her sister's identity. My great-aunt died at an early age, and my grandmother took her sister's identity because her sister had been born in the US. This was sometime back in the day near El Paso. It was also in the days when the border was very easy to cross back and forth. At some point, it became harder to pass between Mexico and the US. My grandma thought it would be best to become an American citizen. So she used her dead sister's identity papers.

As a result, my mom grew up poor, very poor. She picked cotton. She picked fruit. She assisted in doing all kinds of work around the house. She was like a second mother to her siblings. It was a big family, and her parents had to work to feed them all and keep things running. So my mom did the cooking, the cleaning, the ironing of her dad's work clothes, and so forth.

Did she grow up impoverished? When we talk about poverty, there are different thresholds. For example, for her Christmas consisted of getting one used toy. By the time she was eighteen, my mom could see that her future if she stayed in and around El Paso would be pretty limited. She would have to get a job in a warehouse, or a cannery, and live a very simple, very basic lifestyle. There is nothing wrong with that kind of life, but she wanted more for herself.

GROWING UP IN "MEXICAN" LA

My mother's kind of a dreamer. She saw on television the glam of Hollywood and Los Angeles. So she wanted to go to LA to see firsthand what it was all about. At eighteen, with fifteen dollars and a change of clothes, she got on a bus and came to the city. She got work doing ironing, working as a gas station attendant, etc. She was living in an apartment where my uncle, my dad's brother, was the maintenance man, the superintendent. My uncle called a buddy of his and told him that these two new girls had moved into his building. He suggested that his brother, who would become my dad, come and hang out. Maybe the two of them could get to know these two young women. That was how my father and mother were introduced.

My dad's story is slightly different. He's Puerto Rican. The Puerto Rican experience in America is very different from the Mexican experience because Puerto Ricans are born American citizens. I think there are different social pressures, and not so much pressure on Puerto Ricans in the US.

Though my parents, Carlos and Alice Vargas, were from different backgrounds and countries, they had one thing in common: their early lives were rough, especially my dad's.

My dad grew up in the Bronx, a big part of New York's Puerto Rican community. He didn't want to go home and face an alcoholic father who was also abusive. At one point my paternal grandmother escaped that abusive relationship. But my dad stayed a street kid. He got involved with the gangs in LA, Echo Park, the 18th Street area. He got involved in gang fights and got arrested at sixteen or seventeen. This was when the Vietnam War was on. He was presented with a choice: go to jail or enlist in the military. He chose the latter and that's how he got into the Marines.

Despite the family being from two different backgrounds, I didn't fully experience both cultures equally while growing up. The LA area is very Latin, Mexican-heavy. I was raised really only understanding my Mexican culture. I knew I was Puerto Rican but I didn't know any other Puerto Ricans outside of a few family members. I didn't know much about the Puerto Rican culture. I didn't go to Puerto Rico to visit until I was seventeen, so we didn't really have a connection to that family. I had my Puerto Rican grandmother and uncles and cousins, but they too were very LA-indoctrinated. They were more Mexican Latino than Puerto Rican Latino. The Spanish is very different. The foods are different. I grew up eating both foods, but not knowing the difference between the two, I couldn't tell what was Puerto Rican food and what was Mexican, because my mother cooked both. And she cooked them both very well. So in my head, it was all Mexican food.

FACING RACISM FOR THE FIRST TIME

While there was racial kidding around in my childhood, I don't think I ever ran into racism. I had different friend groups in high school, but those weren't based on race but rather on what we did. My baseball friends were predominantly white. My school friends who were in other sports were Hispanic. So to my white friends, I was the "beaner," right? They would joke, "Oh, hey beaner," and not in a derogatory sense. It was more of just finding ways of picking on each other. On the other side of it, to my Mexican friends, because I didn't speak Spanish, I was the white boy. I was the coconut—brown on the outside, white on the inside. I noticed this teasing but it didn't affect me. They were going to talk about me, but I didn't really feel any way about it the way people today get upset about being called certain names.

My parents didn't raise me to see color. My parents never talked about race. They never talked about anything like that. They never made me feel that because of my skin color, I would be judged or treated differently. I never was made to feel that because we're Hispanic, a white person has an advantage over us. I never felt that in my entire life. At the same time, we were never ashamed of our background. My parents mostly spoke Spanish in the household, until we got older and we started to assimilate more to the surrounding world. Eventually we only spoke English to each other. My mom spoke heavily accented English for a while, but over time, that changed too. I guess that shows how much we assimilated.

I am the youngest in the family, with an older brother and sister, and a twin sister who was born two minutes before I was. We're close and we have fond memories of growing up. We were very competitive, very active in sports. We played lots of games at home—dodgeball, basketball, and our own version of the TV show *American Gladiators*. One sister became a teacher, my twin sister is a speech pathologist, and my brother is a firefighter. They all pursued a path of being of service to others.

My father paved the way for us. My dream in life was to play professional baseball, and I played year-round from the time I was seven. That was the most important thing to me. I wasn't drawn into the gang world. I never allowed myself to be, and if I had, my father would have kicked my butt. But we lived in a predominantly white neighborhood. Four or five blocks away was the territory of some pretty well-known street gangs. So I lived close enough—we called it the other side of the freeway—that I had friends who lived over there. If I went over to hang out, my mom was a little more attentive and made sure that I was home by a certain time. If I was in our neighborhood, she was a little more relaxed.

I never knew that racism existed. I never felt like it did. I got in trouble once as a kid and cops pulled me out of a vehicle with guns pointed and all that. It actually seemed justified to me because my friends and I were being idiots. I never felt targeted by cops because of my skin color or felt any other kind of discrimination. It was happening, but I didn't acknowledge it until I got older.

I moved closer to making my sports dream a reality when I went to Kentucky to play college baseball. It was different than in LA. People called the people who cooked and cleared tables and did dishes "amigo." I thought that was kind of crazy. They balled all those people up into one subcategory, and that was that.

People would ask me what I was. I told them Mexican. For a long time, I wouldn't say Puerto Rican because people didn't really get what *Puerto Rican* really meant. I found myself having to defend my answer. People wouldn't believe that I was Mexican. They'd mention the migrant workers in the area, the ones doing dishes in local restaurants. Those were Mexicans. I wasn't one of them. It's hard to express how all of this felt, and what I thought at the time. But I did make it a point to find out the names of those individuals. I wanted to treat them with respect.

★★★★★★★★★★★★★★★★★★★★★★★★★★★★★★★★★★

Vince's dream of playing professional baseball eventually fizzled out. Not unlike many veterans' origin stories, struggling academically and with a newborn daughter to support, he decided to take his older brother's advice and joined the Army. An athlete by nature, and eager to take on a challenge, he enlisted as an infantry soldier with an option to become a Ranger, one of the most elite soldiers in the US Army.

Once he made it through the Ranger Indoctrination Program, he was assigned to 2nd Battalion, 75th Ranger Regiment, and shortly after that he was deployed to Afghanistan. While on that first deployment, he thought about his future. His father was a fireman, and Vince had been taking emergency medical technician (EMT) classes prior to enlisting. During one squad-room conversation about future plans, a staff sergeant, Ricardo Barraza, brought up the Border Patrol. He mentioned that they actually had special operations units within the organization. As Vince recounts in his book, *Borderline*, Barraza said something that really piqued his interest. He claimed that working for the Border Patrol would be a lot like being an Army Ranger—only as a civilian. The idea of being able to transfer his skills was really appealing.

Eventually, after earning his Ranger tab and serving three total combat deployments, Vince decided to get out of the regular Army, but he continued to serve in the US Army Reserves until he retired in 2022.

After leaving active duty, he pursued a career as a firefighter. He admired the profession, but it just wasn't the right fit for him. Reflecting on what Barraza had told him, and wanting to be a provider for his family and make his father and himself proud, he joined Border Patrol. Just getting accepted was an achievement. The training was tough—his class experienced a 40 percent attrition rate, and working in a military-like environment wasn't for everyone. From 2009 to 2015, Vince was a federal agent with the Department of Homeland Security. He understands that even saying "Border Patrol" can produce confusion in a person's mind.

★★★

WHAT IS THE BORDER PATROL, ACTUALLY?

The Border Patrol protects, yes, the United States border—anything outside of a port of entry. Our ports of entry are managed by customs. It's illegal to cross the border at any point other than a port of entry. So I was responsible, along with my colleagues, for detaining anyone who came in and out of this country illegally. We guard the borders across the nation. What people think it is? It isn't. This isn't the typical kind of police work where I engage in a high-speed pursuit, pull them over, take my gun out to get them out of the car, and tell the folks they're under arrest. We carry a gun at our hip but ninety-nine times out of a hundred, that's where it stays. Most of the time, especially as things changed, we'd encounter someone we suspected of having crossed over illegally. We would approach them and have a conversation, asking them if they'd crossed illegally. They'd say yes, and we'd take them in and process them. There were few runners. They just walk across the border and say that they were claiming asylum. Very rarely is the arrest dynamic. There is little to no aggression.

Those times when I did encounter a runner, I'd take off after them, catch them by the shirt or whatever, pull them to the ground, and put zip-ties or cuffs on them. I'd inform them that they were under arrest. No fight. No use of force needs to happen. We load them in the truck, make sure they're not injured or otherwise in need of medical attention. We provide them with water if they need it.

★★★★★★★★★★★★★★★★★★★★★★★★★★★★★★★★★★★

Like many veterans I know, law enforcement officers and federal agents who have really seen life-threatening situations rarely tell their stories in a way you'd expect. They generally downplay the danger, perhaps because it becomes so commonplace, they don't recognize how unique an experience theirs is compared to everyone else's. In Vince's book, he recounts many encounters that were dynamic, to say the least, perhaps not so much in the sense that they involved a violent confrontation, but in that the stakes were very high.

Back in 1998, the US Border Patrol created an initiative to have specific capabilities, and formed a team to conduct special operations. Now that team is part of its Special Operations Group (SOG), formed in 2007. Those capabilities include a police tactical group called the Border Patrol Search, Trauma, and Rescue Unit (BORSTAR) and the Border Patrol Tactical Unit (BORTAC). Vince was trained in tactical emergency services, medical evacuation, and search and rescue. In order to qualify for consideration to join the unit, you had to do two years of service. He volunteered and was accepted into the training program. He did five weeks of initial training, then another six weeks to become certified as a basic EMT, and then another ten-day basic tactical medicine course.

What Sergeant Barraza told Vince proved to be true. As a member of BORSTAR, Vince would be performing tasks similar to what he was doing as a Ranger. The training involved not just the medical component, but also search-and-rescue skills—rescue techniques, land navigation, communications, swift-water rescue, technical rope rescue, and air operations like helicopter rope suspension training.

Their responsibilities don't just focus on those crossing the border; they are sometimes called to assist hikers and motorists, or even help to locate and assist other Border Patrol agents. With a limited number of agencies across the country trained so well for emergency situations they are also called upon outside the footprint of the US border in extreme situations like Hurricane Katrina in 2005. For a combat veteran like Vince, especially with his experiences as an Army Ranger, this felt like a natural calling to serve as a civilian.

★★★★★★★★★★★★★★★★★★★★★★★★★★★★★★★★★★★★★★★

HOW BORDER PATROL RESCUES PEOPLE

I was a medic in Border Patrol's special group dedicated to the purpose: BORSTAR. BORSTAR was necessary because as more and more people try to cross the border, the number of deaths and injuries goes up. Some of the territory that agents cover is pretty remote. So, when somebody gets injured or lost—agents or migrants—they can be a long way from a hospital. They may need immediate intervention. We see a lot of dehydration cases, and sometimes just giving them water isn't enough. So I'm certified to push fluids. I can do IVs. You name it, we can do it.

Again, that kind of work appealed to me, especially as I transitioned out of the military. It required a shift in my mindset. In the military I was an instrument; I was trained to kill high-value targets. In BORSTAR my job was the opposite: saving people. I enjoyed that a lot. I've saved plenty of lives, and I feel very proud of doing that. Some were American citizens; some were illegal immigrants. Their status didn't matter to me. I was risking my life to save theirs, regardless. That's where my heart's at. Our team, we've rescued so many people, and that's just in the border sector team. The Border Patrol as a whole rescues many people. Most people aren't aware of that side of our job.

Critics of the Border Patrol want to demonize us but we're the ones who are there to help. We're a very different kind of law enforcement than most people understand.

One example of what motivated me to join BORSTAR happened just a week or two into my Border Patrol work.

CONSCIENCE, MATURITY, AND THE BIG QUESTIONS

We saw a group that was coming across the Rio Grande. We heard a commotion, a lot of screaming. We saw that one of their guys was struggling to keep his head above water. He was drowning, and he was barely staying afloat right out in the middle of the river. I grew up swimming at the beach. I used to boogie-board, all that stuff. And as I was standing there, I felt compelled to do something. Like I said, I was new on the job, maybe thirty days in. I looked at my senior officer in charge of me and asked, "Hey, man, what do we do?" He told me we couldn't do anything. The group and that guy were too far out for us to try to rescue them. I told him that I could swim really well. I asked if I could at least try. He said, "I don't recommend it."

I couldn't just do nothing, so I told him, "I can't just sit here and watch him drown, bro. I'm gonna go." I took off my gun belt and my pants. I knew I had to lessen my weight so as not to be dragged down. I was down to my underwear and shirt. I knew how cold the water was and needed some insulation. I jumped in and swam over to the guy who was struggling. As I got really close to him, the river's current pulled us onto a sandbar, a very shallow area. He stood up. I stood up. He hadn't broken the law, because technically the border is in the middle of the river, and that's where we wound up. I hadn't crossed over. He hadn't crossed over. So neither of us had broken the law. We just stood there looking at one another. After a short time, I turned around and swam back. Everybody on our squad was surprised that I was willing to risk my life for that dude. They noted that I didn't know who the guy was. He could have been a drug runner. I told them it wasn't about that. I didn't want that man's death on my conscience. Just to watch a human being drown? Who could just let that happen?

I have to believe that incident created a ripple effect. Other agents started to see that there was a bigger picture here than just defending the border. Back then I didn't think about the immigration issue at all. I had my humanity, of course. But I was really focused on the special operations component of the job. What I cared about mostly was that I had a good job that paid well and helped me support my family.

I was stopping drugs from entering the country and potentially stopping a future terrorist attack in America. That was me being a patriot, just like I'd been in the military. As a Border Patrol agent, those special operations things, combating drug smuggling, disrupting cartel operations, that was the comfortable side of the work. I didn't want to think about the immigration aspect because it made me feel uncomfortable. A portion of the work was stopping people from crossing illegally, and that meant ending their opportunity to get to America. I felt like I was killing their dream when I stopped them and took them to jail. That was very rudimentary thinking on my part. And that's because I didn't really have or take the time to think through the entirety of the immigration program and policy.

I later realized that this was different from being deployed. When I was a young soldier, it was about "What do we have to do as a platoon to get home?" In that way, it was like Border Patrol work—I didn't really know about or care about all the implications of what we were doing in Iraq and Afghanistan. What effects our actions had on foreign policy, or how the rest of the world perceived the US—neither topic was on my mind. We didn't think about the deaths we racked up. It was more about keeping us safe and getting home. As I got older, I came to look at things through a bit of a different lens. I understand better that those bad guys we were hoping to kill were fighting a fight that they believed in just as much as we believed in the fight we were fighting. I understand now that the guy on the other side of the rifle is pretty much the same. Consequently, I think I hold less anger and resentment. In the Border Patrol I have every intention of saving every life possible. Thus I don't feel any guilt, or question my path with God.

BEING OF SERVICE

I think that when you train for much of your life to take other human beings' lives, you end up in a different frame of mind. For a long time I questioned my walk with God and whether he would accept me after all the things I was a part of during war. For a long time I wanted to gain "good points." I felt like I was in debt. I had to get out of it by making up for what we'd done. I felt in debt to do right, to do better, to save lives. I was a part of a special operations unit that took lives. You carry that with you at times. There's a lot of collateral damage from war, and you have to carry that with you as well. I think a lot of the guilt was me questioning whether I was worthy.

We rescue more people than any other organization across the nation. Anyone. People die a lot trying to cross the Rio Grande. A lot of them don't realize how deep it is or how strong the current, and so they end up dead. A lot of them can't swim, and they use rafts, floaties, even bags of air. They do whatever they have to do to try to get across. Some find sections of the river that are shallow enough that they can walk across, but even then, they take a misstep and fall in and get swept away and die. There are some sectors that lose close to 150 migrants a year in very harsh conditions.

There's a big spectrum around what we do. We wear many hats. As much as we're law enforcement, we also are humanitarians. Some migrants or others have been out here in the heat for two days. They're dehydrated. Boom, here's your water, here's food. They get our medical attention. We are also crisis response teams. If someone loses their father who drowns while trying to cross the Rio Grande, we're there to console them and hug them. Sometimes I carry kids to my vehicle because their feet are too tired to walk. I'm employed in the most complex law enforcement position in the world.

One time a couple of big rigs were swept away in a flash flood. We got the call and I got the grid location. Eight men had been in those vehicles. We suited up and drove across a large ranch. We got on-scene and saw the men above water, standing on the roof of one of the trucks. The truck had gotten lodged, so it wasn't moving. And we rescued them. You have to keep in mind that when we get these calls, they don't tell us if the people in trouble are American citizens or not. They just say, eight people. And those people need our help. We don't inquire about their ethnicity or whatever. Yes, we're at the border, so they might be illegals, but no one cares about that in this kind of situation. We just go and do our jobs. We risked our lives that day to swim and boat out to where they'd gotten snagged. When we finally got to them, I got a big hug. A few of the guys were almost in tears, saying, "Thank God you're here." I knew that my job was something that I could be proud of. Guys on those kinds of teams are very special. They're willing to risk everything to save a life.

THE COST OF FACING DEATH EVERY DAY

Unfortunately, we sometimes get involved in recovery and not rescue. I remember once, we got a call from a girl who needed help. She called 911 and they triangulated her position out in the desert. What happens when someone is in distress like that . . . they don't stand still. When they're very panicked, they just keep wandering. And she wandered to the point where no one could get to her in time. She died alone, half-naked with the word *HELP* scratched into the ground with a stick nearby. It's not uncommon to see several dead bodies a year in your career. I had a kid drown right in front of me. I couldn't rescue him. I threw the flotation bag but he couldn't get it and disappeared for three days in the river. Later, his body finally came through. So, it's just the nature of that lifestyle.

When you sign up for the job, you know that you're going to see some things. It's never easy. It haunts you in a way. I was a big drinker. I had struggled with post-traumatic stress disorder (PTSD) from combat and used alcohol to help me deal with that. So on Border Patrol, I was piling trauma on top of that and drinking as a way to move through it. I didn't start to heal from all of it until I actually left law enforcement and started working on my wellness. I now have my master's degree in psychology and I'm going toward my PhD. It's something that's really important to me. I want to try to help others in the same capacity as I did for myself.

I've walked a path of trying to heal myself and find answers. I've done that for a lot of years, and I now feel like I'm in a position where I can give a lot of guidance for others. I have a nonprofit organization that focuses on helping veterans and first responders and their families. My family has been affected by my own hardships; my wife and my kids have seen me experience post-traumatic episodes that I had no control over. I want to stay in the medical or healing and helping space to offer guidance to help others heal from their own moral injury and their own post-traumatic stress and their own regrets.

★★★★★★★★★★★★★★★★★★★★★★★★★★★★★★★★★★

> Vince is right. Most people and most Americans who are passionate about the border regardless of the side they fall on have no idea what Border Patrol agents see and respond to every day. People like Vince make it their life's mission to help those most vulnerable among us. As a soldier it was to protect the civilians here at home. But as a Border Patrol agent, it was

oftentimes the innocent children of those who were breaking our laws. Knowing all too well myself what it feels like to not be able to save every innocent life, I still can't imagine how Vince felt after a hard day on that job.

But Vince is a special guy. He hasn't just found a way through it himself, but now helps others do the same. The name of his charity—Beterans—has dual meanings. It celebrates the fact that many veterans do return from service and accomplish many positive things. While he understands that many organizations are being formed to assist veterans and focus on the issues they deal with, including suicide, at times he finds that the emphasis on the negative aspect, on how people are broken by their military service, does a disservice. I completely agree. Veterans aren't just all broken and disordered by their combat experience. Some of us experience what my friend Ken Falke calls a "struggle well" or post-traumatic growth. Much like challenges in sports or demanding occupations that make us wiser and more prepared for the next challenge, trauma, when managed and processed correctly, makes us stronger. Vince wants to celebrate those veterans who achieve success and encourages veterans to talk about how they continue to better themselves and their community. That's the other element of the name Beterans. We should all strive to be better and do better. Based on his experience in struggling and overcoming issues that made his transition difficult, he offers guidance to veterans, and what he says applies to anyone, through that organization. He uses his podcast, *Borderland with Vincent "Rocco" Vargas,"* as another forum in which veterans can hear and participate in discussions about these topics.

MOVING BEYOND BORDER PATROL

What I found over time was that working for the Border Patrol wasn't the defining thing in my life. That's not to say it wasn't valuable work. It was. But one of the reasons I left it after seven years is that when you live the life as part of a special operations team—BORSTAR, the Rangers—you live for the team, and you live in that world. You give one hundred percent to them, and your family gets what's left. I was going through my second divorce, and the second marriage was kind of a fluke, but anyway it was ending, and I was faced with going back and being a single father of four kids. I also knew that with that job, I could give my kids a paycheck but not what they really wanted and deserved.

And to be honest, I was dealing with my own PTSD. When you apprehend drug smugglers who have been carrying dope on their backs for however many hours, they smell like people in Afghanistan. I was operating in a similar terrain to the Middle East, and the smells—burnt trash, sewage kinds of smells—brought me back over there. I was carrying a rifle. I was wearing similar footwear, the same kind of helmet and gloves. I got confused a few times and was wondering if I was on the US-Mexico border or somewhere in Afghanistan. A couple of times, I edged uncomfortably close to engaging on someone incorrectly. I could have taken someone's life, and I would have gone to prison. That's when I knew I was struggling with more than I'd ever been willing or able to admit. Drinking was my coping mechanism, and the excessive nature of that was having a ripple effect throughout the rest of my life and my job.

I also realized that working at the border is very different from other jobs. Everybody understands a firefighter and what they do. Everybody understands the doctor and the nurse and what they do. How many people really understand Border Patrol? Very few people understand that in every situation you come up on, you have to make a decision about which hat you're going to wear. Are you saving a life? Taking a life? Consoling a family member who just lost a loved one? Are you providing medical interventions? Or are you on a humanitarian mission? So, it's very complex. And it's even harder when you are Hispanic, and you are dealing with the idea that you're doing some of these things, the difficult things, as I said before, about taking away their dream from your own people. I didn't feel underappreciated, but I did feel misunderstood. Which is understandable—after all, I also didn't really get the complexities of immigration and identity until I spent those seven years on the borderline.

Another thing people don't realize is the critical role we play in drug interdiction. Some people think the worst of us, and if you believe the media accounts only, it seems that all we do is go after migrants and deny them the opportunity for a better life. Or that we do worse. But along with search and rescue and saving lives, we apprehend more drugs than any other organization or agency in the country. That includes the Drug Enforcement Administration. But you have to understand what I mean by "apprehend." We seize the drugs and the runners, but we don't have a prosecutorial arm at the Border Patrol. Once we catch the bad guy and get the drugs they're trying to smuggle, we have to turn them over to another authority. We have to give what we've seized to some municipality, some county, the FBI, the DEA, whoever. They then take the case through to the prosecution level. So, most often, they claim the seizure as their work. We do some really good things that nobody really hears about, but for the most part that's okay.

What isn't okay is that people make assumptions about what we do, without knowing what it actually is. Many people just see that one thing—arresting migrants—and either hate us for doing it at all, or, in some cases, for not doing it more. I was in a tough position being Hispanic. People who didn't understand all the things we do, particularly those of my ethnicity, viewed me very negatively. I wrote my book hoping to make clear this distinction: there's a big difference between border security and immigration policy. It took me working for the Border Patrol, being Hispanic, and really investigating for myself personally and professionally the issues that I hadn't really faced growing up. I wasn't aware of what brown people like me experienced when being discriminated against. I didn't fully understand all the ways that my racial identity functioned. I'm still working through all that. In fact, I'm working on a paper about that subject for one of my courses.

What I do have greater clarity about is that border security versus immigration policy distinction. I know that just about every American thinks we should have a secure border. But what does that really mean? If you ask someone to tell you what a secure border looks like, they could tell you, but if you ask other people, you could get very different visions. That's not even getting into the question of how we get there.

In the past, I've used people and their homes as a metaphor for border security. We all have assets that we want to protect. We do what we can to lower the risk of those things being taken from us. We also want to feel secure. That's a bit harder to quantify, but if you look at the measures people take—living in a gated community, dead bolts, alarm systems, security patrols—you get a sense of how much risk they're actually willing to live with. And that can change. That means that it is dynamic. It's also relative. What you and I think of as secure or risky will vary.

The same thing is true at the national level, except that there's a difference between deploying personnel, technology, and infrastructure, because those things don't equate to a secure border. They help.

They're a good start. For a long time, San Diego was looked at as the best example of how surveillance measures and other things contributed to it being the model for border security. It has the most fencing, the most walls. But 85 percent of the advanced drug tunnels along the border were located there. How does that fit in with the idea of a "secure" border?

We have to get to a point where we can agree on some level of acceptable risk. We have to balance security and freedom. Those are *big* questions, and it's not just border security but immigration policy.

I get it that members of my community have different views about the role that Border Patrol plays. I didn't appreciate being labeled or attacked the way that I was. I hope people can learn and question and not just blow up with opinions based on not knowing the truth. Yes, I struggled with the question about what I was doing to "my people," but I'm also a patriot. I've demonstrated that in a bunch of different ways and at different times. I don't need to defend myself or my choices. I'm good with them. I've done good things. Like anyone, I hope that before you judge me, you know me.

It makes me sad and a little bit angry that some people say that the Border Patrol is always shooting Mexicans. They're not. There were fewer than two lethal engagements on the entire border in a year. Compare that to the thousands of assaults people commit against agents. That's a much higher number than for any other law enforcement agency. That adds to the risk. Agents need to feel secure too. They're human beings doing a tough job in an environment in which they don't shape policy or the national discussion. But they're out there doing good things to help people—illegal immigrants and others—while still laying down the law.

People often refer to guys like myself and Vince as "having lived many lives" when we tell them our career trajectory. I met Vince when he was an internet celebrity known by the nickname "Rocco" and making hilarious veteran-centric videos and T-shirts, even helping produce, write, and star in a crowdfunded movie, *Range 15*, that was a huge hit among the post-9/11 veterans and Second Amendment communities. As a matter of fact, you may have seen him on in the hit TV show *The Mayans*, where he played Gilberto "Gilly" Lopez.

But the truth is, it's just one life. We get one shot at this, one timeline where every decision we make leads to the next consequence, every sacrifice we make comes at a cost we'll feel for the rest of our time on this earth, and every opportunity to do the right thing is a test of our true character. Whether it was as an Army Ranger, a firefighter, a dad, a husband, a Border Patrol agent, or even an actor, Vince has always wanted to serve others. He has always been willing to sacrifice and he has habitually taken his challenges and even failures as a lesson learned for the next opportunity.

As grateful Americans we often recognize the men and women of our military. We have Memorial Day and Veterans Day parades to acknowledge their sacrifice and service, but do we ever truly acknowledge or understand the service of those protecting us and preserving human life at our borders? They are perhaps our last line of defense when the enemy is at the gates, but do we know the price they pay? The sacrifices they make? Through his platform Vince is helping bridge this divide between the sheep and the sheepdog standing watch. And that's exactly what Vince is, both literally with his past work and figuratively now. Working to tell the story, the truth, he is nothing less than a watchdog. For us.

Do Unto Others

— ★ ★ ★ —

STEVE HENNIGAN
★ LAPD ★

"Treat people how you want to be treated. Look them in the eye, talk to them."

A FIRST ENCOUNTER WITH THE POLICE

In 1973, I was a third grader. Me and a few buddies walked to a park from their nearby apartment buildings and homes in Canoga Park, California. Along the way, we tossed a football in the air.

My dad had impressed on me the value of working hard. In fact, I'd learned the same lesson from my mother. Though my parents divorced when I was only four years old, I mostly had a good relationship with them both. Mom had moved south from Northern California with the kids, uprooting us from friends and family there. But her intention was to make the best of it for us and herself and that meant working long hours as a manager of a men's clothing store. Do an honest day's work for your wages. Don't complain; just get the job done.

So, with those values in place, when a black-and-white LAPD squad car drove up, there was no reason for my heart rate to quicken or my stomach to drop. The two officers were big white dudes, and instead of questioning us, one issued an order: "Hey, go out for a pass."

I did as instructed and ran a pattern, watching as the ball sailed over my head. This guy had an arm on him, as the saying goes. I adjusted my route and speed on the next play and made the catch.

My encounter with the police was a positive one. I was naturally inclined to view police officers that way. I was a good kid, mostly, and never had a reason to think they were there to do anything but serve and protect. On the other hand, my older brother had a different relationship with them.

I can't really explain it, but from the time I was a preschooler, I wanted to be a cop. My grandfather worked as a gas attendant for the Alameda County Sheriff's Office. One of my uncles served as a Sacramento cop. Growing up in the late 1960s and early '70s, I watched the

TV show *Adam-12*, a weekly crime drama that followed two LAPD officers whose call sign was 1-Adam-12. I still remember the show's signature lines of a dispatcher, "1-Adam-12, 1-Adam-12. See the man . . ." that played over the car radio.

Unlike a lot of young people, I never let go of my early answer to the "What do you want to do when you grow up" question. I knew, but I also knew that I was too young to apply when I graduated from high school, knowing you had to be twenty-one years old to do so. While in junior college, I gave a brief thought to joining the Coast Guard. If you had asked me years earlier about choosing a life in the military, I would have laughed and said, "Eff that." The Coast Guard was only taking a few applicants a month, and I didn't like the uncertainty of that. Who knew when I'd get accepted? And waiting for that would mean putting off my dream of becoming a cop for an indeterminate amount of time.

I was trying to build a case for myself to become a law enforcement officer, after all.

That's what all this was about. Well, that and being on academic probation at the junior college. Finally a buddy, a fellow offensive lineman who had joined the Marines, told me that being in the Corps was like being back on the football squad. I was unsure, but late one night, sitting up watching television, the program I was watching was interrupted by a news bulletin. On October 23, 1983, terrorists—suicide bombers—drove two trucks full of explosives into buildings in Beirut, Lebanon. I was like, *You know what? I guess I should go into the Marine Corps now.* It was just like seeing movies back in the day

Despite my dad being a Navy veteran, I didn't want to swab decks. I wanted to be in the infantry, something the Navy doesn't offer. I did have a plan, and I figured that after four years, I'd be twenty-two and likely to be able to start my career in law enforcement.

As someone who enlisted in 2005, I'm a part of what we call a post-9/11 generation. Most of us who fought in both Iraq and Afghanistan joined as a response to seeing the attack on that fateful day in 2001. We often refer to Pearl Harbor as the only similar event that would inspire a generation to sign up to go to war, since the Vietnam conflict was largely fought with draftees who didn't have a choice. But for men and women like Steve, there are other moments in history that served as inspiration. One of them is the Beirut bombings in 1983. Following an Israeli invasion of southern Lebanon to fend off the guerrilla forces of the Palestine Liberation Organization (PLO), a series of escalations resulted in hundreds of United States Marines and French military being stationed in the region to facilitate a peaceful withdrawal. However, in the wake of PLO forces leaving Lebanon, a remnant group of terrorists decided to attack the US and French forces. Two barracks buildings housing American and French service members were bombed as troops lay sleeping on a Sunday morning. In total, 307 people were killed. Two hundred and twenty Marines lost their lives and another 128 were wounded. Eighteen sailors and three soldiers were also killed in action. For the Marines, that was the largest single-day death toll since World War II and the Battle of Iwo Jima. For the US armed forces, that was the largest single-day death toll since the first day of the Tet Offensive in 1968 in the Vietnam War. The group responsible for the Beirut bombing, Islamic Jihad, is considered the first iteration of what is now known as Hezbollah (Party of God), an Iranian-backed group in southern Lebanon currently engaged in conflict with Israel.

As Steve told me, "Damn it, I was pissed." Perhaps for the first time in his young adult life an overwhelming sense of responsibility replaced his personal ambitions. It was no longer about building experience to become a cop; now it was about defending his country. Steve enlisted the next day and was disappointed to find out that he wouldn't go to boot camp until February 1984. He was also wary of what his dad might say. His father had served in the Navy during the Korean War, working as an engineer aboard a munitions supply vessel. Steve was right to be worried. His dad was a little angry with him at first, mostly because of how hasty Steve's decision seemed. I guess Steve knew his dad well, because his initial reaction wasn't necessarily one of support. But he later found out it was just how hastily Steve had decided to join that upset his father, not his motivations.

His mom was more understanding. She sent him a letter before he left for boot camp, telling him that he needed to listen to his drill instructors. He chose to abide by that, and even when he thought what he was being asked to do was BS, he did what he was told and graduated as the Honor Recruit.

He continued to serve with distinction, earning a Navy Achievement Medal for rescuing two civilians from drowning, as well as a Marine of the Year while serving in Puerto Rico. Despite his successful career as a Marine, with seemingly no wars on the horizon, Steve decided it was time to get out and pursue his childhood dream once again. He resisted those around him encouraging him to reenlist, and on October 1, 1987, he drove to Los Angeles to take a written test, a prerequisite screening test to join LAPD.

RESPECT AND THE POLICE

During the oral interview phase, I shared the story of my encounter with the LAPD officers throwing me passes. The interviewers were impressed because I didn't resort to a tried-and-true answer to their question about why I wanted to be a police officer. I knew that the easy answer was that I wanted to help people. I did feel that way, of course, but my reason went deeper than that. I wanted to live a life of purpose. I wanted to demonstrate my integrity and my humanity. And for the next thirty-five and a half years, that was my goal. And, initially, at least, I believed that the public believed that what I was doing was honorable—honorable but complicated in a way the TV shows I watched never were.

This tells you a lot about the times I grew up in. Like I said, I had that positive experience with the LAPD. I also remember how my brother and his friends would hang out in the park. The cops would drive by like usual. And my brother and his friends would all start yelling "Pig! Pig!"

So this one time, I'm at home in the apartment with my mom. In walks my brother. He is in seventh grade or so. And he's crying, and it looks to me like somebody beat the crap out of him. Mom asks him what was going on. He tells her the truth. The cops drove by, they called them pigs, and the officers chased them down and put a beating on them. Back in the 1970s, I hate to say, the LAPD used to own the streets, if you will. That was something that I saw and experienced when I became a young officer. I don't mean that we physically ruled the streets, but we had a reputation, and people used to respect us. All that changed in 1991 with the Rodney King incident.

Steve casually refers to a time when police officers had the full support of their municipalities and citizenry to keep law and order in areas where violence and crime would otherwise flourish. For many officers of that time, the truth was simple: intimidation yields results. A part of me sees the rampant crime in cities across the country today and wishes things were still that way. We certainly don't support police with trust and confidence to keep the streets clean of crime anymore. But like any group of people entrusted with power, it only takes one bad decision to ruin the trust completely. The now-infamous Rodney King incident was especially harmful to that trust. Not just because it was one of the first times blatant excessive force used by police was caught on camera and released to the public, but because none of them were convicted of wrongdoing.

When the verdicts were announced in 1992, a swarm of riots broke out all over Los Angeles. People who were supposedly upset because they believed they were being treated like criminals simply for the color of their skin became just that: violent criminals. Their rampage lasted for six days and needlessly took the lives of sixty-three people. In 2020, we saw a similar powder keg erupted after George Floyd died following his arrest, which was filmed by onlookers. The arresting officer pinned him down with a knee on the back of his neck while Floyd proclaimed he couldn't breathe. That officer was later convicted of murdering Floyd, but in this case, the conviction didn't matter. Before a trial could take place, cities began to burn across the country. More incidents involving police officers using force surfaced almost by the day. Although almost all of the other incidents were found to be justified, rioters masquerading as protestors destroyed public buildings and private businesses, leaving their own communities in ruin for the sake of "justice." No matter how terrible an injustice may seem, we have a justice system for a reason. Any commonsense American knows that the riots of 1992 and 2020 were reprehensible and worsened the relationship between cops and citizens for years to come.

Honorable police officers like Steve acknowledge that those officers crossed the line with Rodney King, and some of those rioters' grievances were warranted. The standards police are expected to live up to are high, and the vast majority of officers live up to them. But it seems, especially these days, that the only headlines come from the few incidents where they don't.

★★★★★★★★★★★★★★★★★★★★★★★★★★★★★★★★★★★

"TREAT PEOPLE HOW YOU WANT TO BE TREATED"

I was still on probation, which is to say, officially working but in my earliest days on the job following graduation from the academy. I was working in the Foothill Division with a senior guy, a great guy, Gary Alvarez. I think at that time he had twenty or more years of experience. One night we were out patrolling, and we got a shots-fired call. Another unit was asking for assistance. I was the passenger officer, and Gary was driving. He was really into tactics, and as we approached, he was telling me how we should handle this. We got to the location and my Marine training took over. We were about a half-block away and I said to him, "Let's not pull on the street." He agreed and said that was what he was thinking too. He came to a stop. We deployed on foot. We started getting to the house and linked up with other officers. A veteran sergeant was on-scene and other officers were spread out. All of a sudden a guy came out of the house and started firing into the air, not directly at us, but still, it was an AK-47.

My Marine Corps training kicked in as an infantryman. I instinctively started shouting to the other guys to get down, to reposition, to get behind something solid, move to this spot or that spot. Then I heard the sergeant, the guy in charge, shouting his orders. I shut up.

I don't think the guy with the AK knew that we were there when he started firing. He surrendered right away. Nobody fired on him, and the whole thing was soon under control.

Well, it wasn't completely over. The sergeant, Wilkerson (we later became good friends), walked up to me and said, "Hey, boot, what the eff do you think you are doing?"

I apologized for jumping the gun and explained about reverting to Marine Corps days. Still, I had no business telling anyone what they should do on the scene. The field sergeant reminded me again that I was just a "boot"—an inexperienced, probationary officer. My partner also let me know that I had done well, but that he "needed me to remember who he was."

Over the course of my career, I learned a lot of lessons. For a number of years I rode a motorcycle. As a motor officer, my primary responsibility was traffic enforcement and safety. I believed that there was the letter of the law and the spirit of the law. I probably issued fewer tickets than the number of people I stopped. I used discretion in dealing with petty, victimless crimes.

It comes back to something my dad told me when I was a kid: you always respect people. Treat people how you want to be treated. Look them in the eye, talk to them. In this line of work, meeting and coming in contact with bad people is a big part of it. I always showed respect to people until they didn't give respect to me; then I would get on their level. So, if they were going to be assholes, I'd still be the professional, but I'd be a professional asshole.

But for the most part, I tried to be as empathetic as possible.

One time, as a patrol officer, I got a domestic violence call. I showed up and was met at the door by an older black woman. She was in her late sixties. We identified ourselves and she told us to come right in. We asked her if she was okay and where her husband was. She pointed and said, "Right there." Over on the living room floor was her husband, dead. He had three gunshots in his chest. *Holy shit.*

I ask the woman, "Where's the gun, ma'am?" She goes, "Oh, it's back in my bedroom." She was as calm as could be, and so was I. I didn't even handcuff her at that point. I had worked with the homicide unit before. I was filling in for a guy in the Wilshire Division who was on vacation. I had interviewed suspects, so I knew how to handle this. I kept talking to her, allowing her to make spontaneous statements.

I also and honestly wanted to be sure that she was really okay. There'd be time to read her her rights and all that. She told me that her husband had been beating her. I could see the bruises on her face and her arms. I told my partner to stay with the body and I asked her to show me where the gun was. As we went upstairs, she told me what had been going on. Her husband had been upset and kept saying that he wasn't the father of their twin daughters. They had been married for forty years, and he would go off on her about his suspicions and frequently beat her throughout their marriage.

She also told me that she was a nurse. I was already feeling sympathetic toward her. She had been a victim of domestic violence all these years, and she was a nurse. I'd met my wife while on duty at a hospital. She was an ER nurse. I know what they have to go through.

She told me where the gun was and we returned downstairs. The scene was being processed and the homicide detectives decided that since they knew me—we'd worked together—and since I'd already been speaking with her, I should continue to question the woman.

She fully admitted that she was being beaten again, and that she'd had enough. She got away from her husband, got the gun, and fired at him three times from close range. As she was being led away after being placed under arrest, she stopped and paused to say to me, "See that closet? Open that door. That's where he beats me. He had it soundproofed so nobody could hear me screaming."

I looked, and verified that what I had heard, which sounds like something out of a horror movie or a disturbing psychological thriller, was true. The closet was lined with heavy wood panels.

All I could think was *Holy shit, the things people will do to each other.* I sympathized with her and could imagine, after having seen so many domestic violence cases in my career and her most recent bruises, what she had endured. How could any man do this to his wife, let alone a woman, or any other person?

I was tasked with accompanying her to get booked. I said, "Okay, no problem." She'd killed someone, but she'd never been in any legal trouble before. She was a professional woman, a nurse, and now she was going to be booked and housed in jail. Everything but the beatings was a new experience for her. So I told her what to expect from all the proceedings of the paperwork and things to what she could expect and how she should protect herself from the others in the holding cell while in the lockup. Also, she had to be photographed to preserve the evidence of the mitigating circumstances. The female officer who did this came back out of the room looking really shaken. She told me that her whole body was covered in bruises. I felt even worse for that lady.

As officers, when you arrest somebody, you're always in fear that somebody is going to try to get revenge on you. We had LAPD officers followed home and shot and killed by someone—the person they arrested or someone associated with them. So there's always that fear you take home.

That's why I reacted the way I did to what happened next. As part of the legal proceedings against the woman, I was subpoenaed to appear in court at a preliminary hearing. I walked into the courtroom and that familiar trepidation was right there with me. As I was walking toward the courtroom, I saw two black women seated on a bench just outside it. They looked alike, and I knew they must be the twins, the ones their father used as a feeble excuse to beat their mother. One of the twins pointed at me and my central nervous system went on high alert.

I heard one of the pair say, "That's him. That's the officer." The other stood up and ran toward me. I tensed, wondering what I'd gotten myself into. But my cop sense let me know after an instant that this wasn't an angry person coming toward me. Soon I was embraced in a hug by the two women. They thanked me. Their mother had told them that I had treated her kindly and respectfully. They wished that we could have all met under different circumstances.

Six months later, I received a radio request to come to the watch commander's office. That was seldom, if ever, a good thing. It was the equivalent of having your name announced over the PA at school to go to the principal's office. I wondered what I'd done. When I was finally let into my superior's office, my mind was clear. A large cake sat on the desk, and the watch commander was smiling. He handed me a card. It was another thank-you, this time from the woman herself. She thanked me personally for how I'd treated her that terrible night. She hoped I would enjoy the cake.

This was a shots-fired call, domestic violence. You go on enough of those you've got a list of possible things that could be going on, and almost none of them are good. I could have gone in there with gun drawn and made this lady lie down in the prone position, and handcuffed her and led her out to the patrol car and thrown her in and taken her off to jail. I didn't do that, and in all my years of arresting people, I hardly ever had to use that kind of force. In my thirty-five and a half years, I might have used force to effect an arrest maybe four times. With all the others that I arrested through the years, I was able to show them respect and talk to them calmly and make an arrest without using any force. Those three or four instances? I had to struggle to get them to submit to being placed in handcuffs. I think that's pretty amazing. I owe that to how I was raised, and what my parents taught me about respecting people.

The woman I'd arrested for shooting her husband had baked that cake herself. She was able to because the charges against her had been dropped. The spirit of the law overrode the black-and-white letter of the law. I'm glad that the system had worked as it should, rather than blindly following the statutes by the book without regard for the circumstances.

I remain grateful that I never had to fire my weapon other than on the range.

In 1994, I was serving as a motor officer. I arrived at the scene of a multiple-vehicle accident in South Central Los Angeles. Surrounding the accident scene were burned-out husks of buildings that had been torched during the rioting. I went to work, taking down information from the drivers, still wearing my helmet. At one point I heard the snap of a bullet going just past my head.

I reverted to training. I didn't duck. I didn't cower. I just immediately drew my weapon. I'd gotten hit with concrete from the bullet impacting the building behind me, just a couple of inches above my head. I advanced toward where the gunshot came from. In front of me was a tow truck. A black lady sat in the passenger seat. The window was down and she had a gun in her hand. I kept advancing, telling her to drop the gun! Drop the gun! She hesitated at first. I was carrying a Beretta, and with that weapon I could go to the half-cocked position on the trigger since it's double action. She dropped the gun and I put out a help call on my radio. I ordered her out of the truck and got her to lie on the street. Extra units arrived. We got her handcuffed. I was pissed-off. I'm not going to lie. She had almost killed me! It turned out that she was an ex-con and was later convicted of attempted murder.

I'll admit that I did keep a cool, calm demeanor throughout the incident. I guess my professionalism showed in not firing on her immediately, but I was a millisecond away. Given all the circumstances, I would have been justified in firing on her. But she gave up.

There was another time when, late at night, my family and I were victims of a home invasion. Alerted by a neighbor, I went into action. Again, I could have fired on the man, who, as it turned out later, was armed but had set the weapon down an instant before I confronted him and subdued him.

I had my weapon drawn and half-cocked again. I kept screaming at him to get his hands up. I went back to what I'd learned at the academy. I told him to get on the ground, shouting at him. He kept telling me that everything was okay and that he was just looking for someone. He said that he didn't have a gun.

Eventually I got the guy down on the ground. All I kept thinking about was that we'd just remodeled the house. If I shot the guy, the round could have gone through him and into a freshly painted wall and he would have bled all over the new carpet.

California law at that time would have been on my side if I'd fired and killed the man. I eventually learned that "the man" was a violent gang member who just twenty minutes earlier had attempted to rape a woman he'd been out on a date with. He was eventually convicted, thanks to me getting the very frightened young woman to testify against her attacker. She feared retribution, but stood up and did the right thing. The man was eventually deported.

So much of Steve's experience goes back to one thing: his decision-making in a moment of stress. I don't think we, as private citizens, truly understand all the things a police officer considers when responding to a call. In every instance that Steve outlines, there's the way he was trained to respond, and then there's the discretion he felt empowered to use. When he walked into a home where a lady had just shot and killed her husband, he had every right to treat her as a danger and bad actor and "let the investigation sort it out." But Steve's intuition kicked in and he felt genuine empathy for her. In this country we all have the constitutional right to protect ourselves from imminent harm, and the default standard shouldn't be that we are automatically treated like violent criminals when we do. But from the officer's perspective, without much information to go on, it takes a willingness to risk his or her own life to treat us with respect as they sort through what just happened.

I can't say I would have responded the way Steve did. I legally carry a loaded weapon everywhere I go. You can bet if someone shoots at me, we're not having a conversation before I shoot back. And you can absolutely count on the fact that if someone breaks into my house, they're going to leave with new holes punched from a 9-millimeter or twelve-gauge. But I also don't have Steve's training or experience. The fact is, he made the decision early in his career to use discretion, to take a little more risk to his own safety to potentially preserve more life, even for those who intend to do him harm. That's the true heroism he exercised for thirty-five years as a police officer and it's the type of sacrifice officers make every day that we hear nothing about.

WORKING FOR THE BOMB SQUAD

The bomb squad was something I'd always been interested in joining. I get it from my dad—choosing big goals to drive for. Always on to something bigger. I ended up at the Redstone Arsenal in Huntsville, Alabama. I was there for five weeks to attend Hazardous Devices School. I spent five weeks in Huntsville studying nuclear, biological, and chemical (NBC) devices and the symptoms of possible exposure to those three types of devices and substances.

I compare that first portion to an abbreviated hazardous materials course. There was a lot of classroom work, overnight reading and studying, and then tests. Eventually we engaged in exercises during which we put into practice what we'd learned in the classroom. The pattern repeated itself: lectures, reading, written tests, exercises.

I was able to recall similar training during my time in the Marines. From that training in January 2000, I was a member of the bomb squad for the next twenty-three years.

It was an escalation in my exposure to danger—I'd gone from being a patrol officer to riding a motorcycle to exposing myself to seriously dangerous substances and explosive devices. I credit my wife for her support. As an emergency room nurse, she was used to seeing trauma cases. She wasn't immune to the stress that that kind of work creates, but she was aware that I wouldn't be happy unless I was doing what I loved. And of all the areas that I worked in, the bomb squad was where I really felt most at home.

When I first started, we were getting a lot of IEDs. We had pipe bombs. I remember one early call about a suspicious bag that had been placed in front of liquor store. At six in the morning, one of the liquor store employees showed up on-site. He found a smoking paper bag. He opened it and saw that a portion of a candle was inside. He also saw a three-sided box that contained a three-inch-diameter, six-inch-long pipe bomb.

The man was fortunate. The candle's wick served as a fuse. In tearing the bag open, the employee had allowed in enough air all at once and caused the flame to go out. Somebody was looking down on him that day, because he got there at the right time and had the right thing happen for him to survive. The guy threw the device into the parking lot, and that's where I discovered it when I showed up to dispose of it.

That was early on in my days with the bomb squad and that was a reality check. I remember being suited up and walking with the disruptor tool. I was talking to myself saying that this is real, dude. You have to do everything you were taught and do it right. If I did, everything would be okay. Once the disruptor tool was in place, I retreated to a safe area. Only then did I realize that a TV news chopper was flying above us. No pressure to have your work being broadcast like that. Fortunately, everything went well. The disruptor went off and the pipe bomb exploded. Debris flew all over the place. Good thing the area had been cleared and that guy had found it and reported when he did. A few parked cars took a hit, but that's a small price to pay.

Our team's work wasn't done. The post-explosion work was just starting. Having detonated the device, we searched for all the pieces we could locate, took photos, otherwise documented all the evidence, and then, with all the bits and pieces collected, performed a microscopic analysis. In this case, the filler used was black powder. We then turned over all the evidence and analysis we'd completed to the detectives assigned to these kinds of crimes. They're called the Criminal Conspiracy Section (CCS). Those CCS detectives take over and pursue arresting whoever constructed the bomb and placed it. In this case, they tied the device to the Russian mafia, which was operating in the East Hollywood area.

Over the years, I had to deal with many types of explosive

devices—from ones to fit inside a single cigarette, a cigarette carton, to much larger. Explosive devices are tricky enough to deal with as it is, but the *I* in IED, their improvised nature, makes them especially tough. I started my career in the bomb squad pre-9/11 and the global war on terror but eventually my two careers—the military and law enforcement—merged. Once we were in Iraq and Afghanistan and IEDs were wreaking havoc on our personnel, the Marines and the Navy sent some of their explosive ordnance disposal (EOD) team members to learn from the LAPD.

At the start, that EOD team came to us hoping that what we were experiencing in LA would give them some sense of what to expect. Unfortunately, the kind of warfare and devices they would encounter were orders of magnitude worse. A lot of the calls we went on were more of a nuisance in comparison. Most of them were vandalism. A few times there was more of a malicious intent—someone trying to kill another person. A lot of the time, those efforts failed for one reason or another, and we were there as evidence collectors. A lot of times, the builders were just inept.

Also, as the war on terror kicked into higher gear here, we got a lot of calls about threats to the many consulates in LA. We got a lot of calls post-9/11 about suspicious items. We'd have to go there and X-ray them. A few times we came to check out dead bodies that had perhaps had something on them, and the suspicion was that they might be suicide bombers. Once we had to X-ray the briefcase of a bad guy who died on the steps of a post office. The suspicion was that he might have been a suicide bomber. Weirdly, though, the number of calls we went out on declined after 9/11. We used to seize a lot of illegal fireworks, but those calls dropped off substantially. The seriousness of the calls ramped up, though. In 2000, we had about 1,000 bomb calls throughout the city for all kinds of things. After 9/11, everybody was suspicious about everything. I remember immediately after 9/11 we had 365 planes at Los Angeles International Airport that had to be searched before they'd be allowed to fly. They'd be towed to a safe area and sit there for a couple of days. We had to physically board the plane, accompanied by a bomb-sniffing dog and its handler. We just didn't have enough staff to get to all of those searches quickly. After that 9/11 bump, the decline began, and by the time I retired in November 2023, we were down to 250 to 300 calls per year.

★★★★★★★★★★★★★★★★★★★★★★★★★★★★★★★

Hearing Steve talk about the differences in domestic bomb threats he encountered and what we saw in Iraq and Afghanistan is humbling. Certainly, we had our work cut out for us with the size and construction of the IEDs we were experiencing. But don't let him sell his work short. It's indisputable that the tactics and knowledge that allowed us to be successful in war were first proven by techs like Steve doing the work at home. And honestly, I can't imagine what it's like to see an IED in my own town that could hurt innocent civilians, people I call friends and family. But that was Steve's experience for more than two decades.

When Steve joined the LAPD bomb squad, there were sixteen members. Eventually that number increased to twenty-eight at its peak. Keep in mind, that's for a city with a population of almost 13 million people. The calls they received on the bomb squad varied greatly—Steve went out on calls up in the hills where a foundation for a new home was being dug, and dynamite from 1800s mining operations were being used. Anytime a high-profile politician came to town—and they came to town a lot—Steve and other members of the squad did what are called find-or-function searches and coordinated with the Secret Service.

★★★★★★★★★★★★★★★★★★★★★★★★★★★★★★★

GIVING BACK

In a varied and very successful twenty-three-year run with the bomb squad, I had done just about everything. I'd also been with a lot of different guys (for the most part), many of whom were fellow Marines. I appreciated the brotherhood I enjoyed as a Marine, as a law enforcement officer, and as a bomb squad member. I understood, but wasn't happy about and didn't participate in, the kind of careerism and backstabbing that came with being in a large organization, with some members who put their own interests ahead of their brothers. Instead of complaining about it, I thought it best to take that attitude on myself.

I found my own way to counter the damaged unit dynamics that I so treasured as a Marine. I did this by acting as a liaison between the LAPD bomb squad and the Marine EOD units. I made the connections, organized a training curriculum that worked in concert with the Marines, and built a program that was mutually beneficial to each group. I appreciated that the young Marines didn't see me as an outsider or an "old dude," instead, on their own or because Marine leadership encouraged them to, they treated me as a valuable resource. Because I'd been at it for so long, and had learned a lot on the job, and on my own, they saw that I had a lot to offer.

Now that I'm retired, I'm concerned that the relationships I built between the police and the military might deteriorate. That saddens me, but I'm glad that at the core of my life, I still have the Corps. I've helped a bunch of Marine EOD guys who've gotten out get jobs with the LAPD and other agencies. I hope they'll foster the same kind of brotherhood and connection that means so much to me.

I credit that connection with helping me avoid some of the pitfalls that come with doing the kind of work I did for more than thirty-five years—the burnout, the mental health struggles with depression, anxiety, and suicide that have become far too common in law enforcement and elsewhere.

Also, I'm a talker. I'll talk to people about stuff that a lot of them don't. I had a friend, Joshua Cullins, who was LAPD. I met him in 2005 and in 2007, when he was in Iraq. They got in this big firefight. Suicide bombers, and all this other shit. He had to put down a few bad guys. One of his guys, right next to him, got shot. He called me on the sat phone right after so he could talk through things with me. I mean, I'm no psychologist. And I know that on the force, guys have the opportunity sometimes to speak with the department psychologist after they have been in an officer-involved shooting. And some did that, others didn't. What got me, though, was . . . that was it. A lot of them didn't speak about it with anyone else. And nobody else approached them. Well, I would do that—both parts. I would talk with others about what I was dealing with and ask others what was going on with them.

I know enough to know what compartmentalization is, but I didn't do that. I didn't think that was healthy. And I thought that the best thing I could do for other people was to just listen. When Josh called me from Iraq, I listened to him. He didn't want my analysis or advice or for me to share my own stories. He just wanted to share what he'd seen and done. And that's all I expected or wanted from someone I went to.

Josh Cullins was killed in Afghanistan. I know the two guys who were with him when he was killed. And every year, on the anniversary of my buddy's death, I call the two of them. I know that they're thinking about it. I know that it's tough on them. But I call them or text them. I have to check in. With other guys, I don't feel the need to wait for an occasion like that. I just check in—cops, veterans, whoever. I don't compartmentalize. I don't drink alcohol to excess. I don't numb it. I don't stuff it down inside me.

I wish I could bottle whatever has helped sustain me and kept me from developing PTSD. I know many cops who have, veterans who have. I encourage friends, colleagues, and even strangers to just get it out and not look at talking as a sign of weakness. Don't rely on the excuse of "No one can relate to what I've been through except somebody who's actually been through it."

Anybody can be a sounding board. And I get it that some people aren't good listeners. So, it does cut both ways. We've got to make ourselves more available to people who we know are in a high-risk group. You don't have to have gone out on a call to a meth lab or shoot-out to be of value to someone who's struggling. I've had guys who saw a lot of combat talk with me and ask questions about what I did as a cop. They're surprised at the variety of stuff we're called on to do. They didn't quite get that we were enforcers, social workers, and dispute-resolution guys.

★★★★★★★★★★★★★★★★★★★★★★★★★★★★★★★

The job of a police officer is about as stressful as it gets. You not only go to work every day knowing you're going to see and respond to the worst individuals of our society, but you'll also see and respond to the innocent victims they prey on. Along with those stresses every cop endures are the societal stresses in today's cultural and political climate. There's the constant worry that someone will film a fraction of what you do in a situation and use that to smear you, or worse, accuse you of committing a crime yourself. In Steve's case, those stresses weren't enough to steer him away from doing the work we all need to be done. On top of all that, he even chose to pursue an even more dangerous specialty in becoming a member of his department's bomb squad.

In talking about the mental and psychological effects of

his career, you might think Steve made it through unscathed, that he was somehow able to see terrible things by day and hang those emotions and memories up on his way through the door at night. But that couldn't be further from what he's saying. The truth is, first responders, much like combat veterans, experience severe physiological trauma, and Steve is no exception. However, whether or not that trauma becomes a lasting wound varies in each of us. And in Steve's case he exercised insightful techniques to help him heal quickly from the things he experienced. He talked to people about what he was seeing and feeling rather than stuffing it all down. He took accountability for his own actions and understood he couldn't control the actions of others. He sought grace and empathy over vengeance and indifference and chose not to self-medicate with harmful vices like alcohol. Whether he knew it or not, all those small decisions led to his ability to have a long yet tumultuous career in law enforcement and come out on the other side with clear eyes and mentally healthy.

★★★★★★★★★★★★★★★★★★★★★★★★★★★★★★★★★★★★★★★

America's Sheriff

MARK LAMB

"We need leaders who stand up
for their officers and stand in
front of them when the bullets
are flying. If you do that, you
make it easier for those under
your leadership to find the
less desirable parts of the job
a little bit more doable."

THE PHYSICAL EFFECTS OF A POLICE JOB

I couldn't afford to live off a police salary alone. I have a wife and five kids to support. So, for many of my early years in law enforcement, I had a pest control business. I would work a twelve-hour graveyard shift. I'd work 6 p.m. to 6 a.m., get off, go home, and sleep starting at 7 a.m. After getting two to three hours of sleep, I'd be back up by 9 a.m. or 10 a.m. at the latest. I would go out and do pest control all day long. And then I would go and get in a workout because that's the only time I had to do it. Then I was back on to another shift.

You try to maintain your body, but it's really difficult. The deprivation of sleep is not good for your body. And because you're physically down a bit, that takes a bit of a mental toll. You've got to be careful you don't become too irritable, angry, or even despondent. In many cases when you're physically tired, it's really difficult to pay attention at home because you have to save everything you're doing for work.

I used to tell my wife, "Honey, don't call me with any minutiae from home. Only call if it's a major emergency. Don't call me, because I don't have what it takes to get into the game and then pull myself out of the game to listen to talk about a bill that needs to be paid and then try to get back in the game. It's too dangerous." At work, I have to be fully present. If, as a police officer, you're not on your game, one split second can cost you your life.

I spent a lot of my years as a gang and drug detective, and most of the time we worked "graves." Starting at six at night until two or three in the morning, four in the morning. So you're working these shifts, and when you're working, you're on a heightened sense of alert. You're expending a lot of adrenaline and you're getting a lot of adrenaline dumps. On top of that, most of the time you're not getting a chance to really sleep and rest and restore your body. That's the physical piece. That's why a lot of these cops are on the fat side. They're strong, they're tough, but their bodies are not where they should be. You're pounding caffeine constantly to make sure you stay awake. You're body's releasing cortisol, the stress hormone, into your system and that's not good for you at all. But eventually your body adjusts a bit, which is both good and bad.

Many cops, after doing this to their bodies for so many years, just become numb to what they see. You don't get those adrenaline dumps as much anymore. You're responding to a burglary or domestic violence and you're as cool as a cucumber because you've been through it too many times. That's because your physical responses are so messed up.

★★★★★★★★★★★★★★★★★★★★★★★★★★★★★★★★★★★★★★

Sheriff Mark Lamb has developed a legion of followers on social media and elsewhere. Most may know him from his appearances on the A&E network's show *Live PD*, but I first saw him on social media pulling people over with a smile to tell them to get out of the fast lane if they're going slow. Commonsense policing . . . what a concept! After following him online, I had the opportunity to interview him live on set for *Fox & Friends*. His passion and demeanor are everything you'd want to see from a leader in law enforcement. And it doesn't hurt that he stands well above six feet tall and effortlessly sports a cowboy hat. He gives off a humble confidence that makes the word *lawman* come to mind.

As Mark said matter-of-factly, "Most people don't really see a cop. They see the uniform, the badge, the gun. They don't see the person." The physical and mental toll shift work takes on an officer's body is something most of us never think about. We may make a joke about their affinity to donuts, but we never consider that such a stereotype may come from their exhaustive hours, high-stress shifts, and a need for just enough sugar to keep their brains and bodies functioning.

Although he believes an impressive aesthetic is an important part of commanding authority, Mark is much more than just an image. He's a dedicated public servant with an acute commitment to the Constitution and of course to the people he serves. His leadership in the arena of politics is more like that of a business leader or entrepreneur. It's his innovative mindset and the experiences he draws from being a later-to-join police officer that give him a unique and valuable perspective.

Of course, having a dad who's also a businessman helped in developing his outlook. His father also had a thirst for travel and immersing himself in other cultures. Those early years with his dad helped shape Mark's views and deepened his appreciation for the absolute greatness that is our America.

★★★★★★★★★★★★★★★★★★★★★★★★★★★★★★★★★★★★★★★

A TRIP TO PANAMA

In December 1989, my mom, my sister, and I flew to Panama to be with my dad for Christmas. Things between the US government and Panama had grown tense. Just after we got there, a US Marine was killed by a member of the Panamanian Defense Force. Shortly after that, another US Marine was assaulted, and his wife was sexually harassed while outside the US-controlled Canal Zone.

On December 20, things came to a head. I was asleep and my mom woke me up shouting about the bombings. From our windows, I could see tracers and bursts of fire across the bay. I could hear the explosions and the machine-gun fire. This went on for hours. Every time things quieted down and we thought we might be able to get some sleep, they kicked up again.

My dad left about thirty minutes into this ordeal but now was unable to come home. Some of Panamanian president Manuel Noriega's "Dignity Battalion" had arrived outside our building in several cars. From the back of one of those vehicles, a black limousine, a man pulled out a rocket-propelled-grenade launcher. We later learned that Dignity Battalion was working from a list of Americans in our building and had coerced the security man in our building to confirm our exact location. The thugs kidnapped one American on the fourth floor, suspecting him of being CIA, and threw him in one of the cars.

Maybe we were lucky, or maybe God intervened. When the Dignity Battalion came back, our building lost power. They were too lazy to climb the stairs. The one American they took was found, sadly, a few days later in a ditch. He'd been shot multiple times.

The next few days were chaotic. The Panamanian police and military had been decimated by the US military invasion. Security was lax or nonexistent. Firefights broke out all over, and assaults and looting were commonplace. Several of the residents in our building had weapons. I was seventeen and a senior in high school. I was assigned a security watch from midnight to six, my earliest graveyard shift, I think. We had a rudimentary communications setup—handheld radios. We'd get reports that the looters or the Dignity Battalion were heading toward our position. Imagine the fear and thoughts going through my young brain and body.

Fortunately, those were all false alarms. But you couldn't tell my body and mind that they were false. My response was real. So was the even deeper appreciation I had for my country. That Christmas, with no gas to cook with, we had cold beans. The grocery store shelves were empty for days. We made it out, and it was an ordeal, and when we came through customs, the banner "Welcome to America" never felt so good before. Later, after serving a search mission in Buenos Aires, Argentina, and having a completely different experience than that Christmas in Panama—the lovely people, the language, the food (best beef in the world!), and all the rest—I was still looking forward to being back home. And, again, hearing a US customs officer say "Welcome to America" was just the best.

I was raised to have a profound love and appreciation for America. I love the freedoms that God affords us and that our Constitution protects. I love and appreciate the Founding Fathers and many others for their contributions and sacrifices so that I can enjoy this country and its freedoms.

★★★★★★★★★★★★★★★★★★★★★★★★★★★★★★

> You might have already drawn a fairly straight line from Mark's extraordinary experiences defending his family as a child to serving his community in law enforcement. But like the rest of us, Mark's path winds and bends and "took the long way around." Instead of connecting the dots and rushing into a military or policing career, Mark enjoyed some of the more unique freedoms that this capitalist country has to offer and went into business for himself, owning and operating a brick-and-mortar paintball store, no less. Mark's father had told him, during the days he spent in Panama, one of life's important lessons: "Sometimes you eat the bear; sometimes the bear eats you." Failure and setbacks are going to be a part of life. Like Rocky famously told his son in *Rocky Balboa*, "It's not about how hard you hit. It's about how hard you can get hit and keep moving forward. How much you can take and keep moving forward."
>
> Like many small business owners of the time, Mark saw the rise in online stores beginning to dominate the market. His little shop in Payson, Utah, struggled to compete. And when a big-box store opened, the decline turned into a downward spiral. He had to close the store and declare bankruptcy.

With no income and no place to stay, he and his wife and five kids moved back to Arizona to live with his recently divorced mother in her condominium. A man with pride in his work ethic and traditional values of "provide and protect" inherent to his self-esteem, he found himself at an all-time low. The proverbial bear was gnawing, but Mark doesn't give up without a fight and he was hungry too. He worked on a new business venture with a cousin. Things were still very unsettled and difficult, and Mark's attitude grew poor.

He knew deep down that he had to shift his perspective and appreciate what he had and not mourn what was lost. Mark is a big fan of these lines from the English poet Rudyard Kipling's classic poem "If": "If you can meet with Triumph and Disaster and treat those two imposters just the same." The poem ends with the finish to that thought, "you'll be a Man, my son."

I don't think Mark had a sense of how disaster and triumph would play such a part in his life and the lives of other first responders. In their lives, more than in many other peoples, the pendulum swings between those two points, creating a true dichotomy of ups and downs, goods and bads, highs and lows.

I'm not sure if Mark saw the full extent of the symbolism, but I do know he eventually saw the light, and of course it had to come on a night shift.

ARMED WITH COURAGE AND A FLASHLIGHT

I had my business, and it was going okay, but I knew I needed something different in my life. Police work wasn't even remotely on my radar. My neighbor was a cop, and he talked me into going on a ride-along. It was a graveyard shift, and it was on an Indian reservation.

We got a call from a dad who found a twenty-year-old male with his fourteen-year-old daughter. So the dad scuffled with the guy. The guy fled out of the back of the house. We show up on the reservation. The house we got called to sat on a patch of land with nothing but desert beyond it. Well, not exactly nothing. The area was dotted with a few abandoned travel trailers.

So, there I was, armed with a flashlight and courage, walking through the desert expanse. We came up on one of those trailers. I walked over to it and peered in through a window. There, among the clothes and trash and debris, I saw what I thought was a quarter-inch-size bit of skin. I immediately thought it was that twenty-year-old in there. So the cops went in and started moving stuff around. Sure enough, he was there. A scuffle ensued and they subdued him after tasing him. They put him in cuffs.

I was hooked that first night. I went home that morning, woke up my wife, and said, "Honey, I'm going to be a cop." Six months later, I was in the academy and on my way to being a cop.

★★★★★★★★★★★★★★★★★★★★★★★★★★★★★

"Armed with a flashlight and courage." What a moment for a man who'd seen such combative things as a child. In the Marine Corps we used to say of combat, "Anyone can be brave. You can hype yourself up and charge into a dangerous situation on adrenaline and ignorance, but courage, courage is acknowledging the danger ahead, the fear it exerts in you, and having the dedication and commitment to run to battle anyway."

Being brave is a moment in time, while courage is an acceptance of sacrifice to something greater than yourself. I don't know if Mark has ever had it broken down this way for him, but I have to believe his use of the word *courage* as some sort of armor or tool is indicative of the fact that his earlier experiences informed him of what real dangers lurked that night, but something needed to be done and he was the one there to do it. He had the courage to do the task. We call this "standing in the gap." It refers to the empty space between the innocent and the evil, the physical and metaphorical time and distance between "everything is okay" and "terrible things are happening." That's firmly where Mark put himself more than a few times in his pre-police life and exactly where law enforcement officers and first responders stand every day.

STARTING AS A COP LATER IN LIFE

I'm grateful that I was raised in a family and in a church that instilled in me the value of being of service. I didn't know that would lead me to police work, but once I discovered the thrill and satisfaction of doing that work and seeing a positive outcome like that, I was hooked. There would be no looking back—except to be grateful that I had made this decision at the age of thirty-four. I was in a very different state of life than a lot of the others in the academy. I'd failed and succeeded. And now, having gone through what I had, making this choice seemed easy. It almost seemed like police work had chosen me.

In most ways, the academy and the work itself came easier to me than it did to those younger guys. I watched them try to respond to calls and make decisions and they struggled. They didn't bring to the table the problem-solving skills that come with living more and more years and trying and failing and trying and succeeding. I entered into this profession with that advantage. But I was very different from my "peers" at the academy and later as a working cop. I don't hang out with cops to this day. It's just not my thing. Don't get me wrong, I bleed blue. I'm one of the guys who loves police work. I don't grumble about it and talk about how much I hate it. I love it. But I think that's because I can look beyond the every day to what being a cop represents: the founding fathers, the Constitution, and its preamble; the very first thing they were concerned with was establishing justice. I think that's what makes America different; that's what makes America great. I think that justice is the backbone of America.

Having that in mind from the outset, being a cop wasn't about carrying a badge or fulfilling a lifelong dream. And maybe, for some cops, that dream bumps hard into reality, and guys get soured. So, when I became a cop, it was easy for me to not fall into that trap of the ideal not living up to the real. The ideal was the defense of the Constitution and justice. That never changed. That's impermeable. The saying goes that cops hate the way things are and they hate change. Well, for me, my patriotism, and my desire to honor my constitutional responsibilities, those won't ever change. And that's a good thing. And sometimes, to achieve that higher aim to keep alive the things you love, you can't hate change.

I've never been the everybody-and-everything-sucks kind of person. I've always been optimistic. I want to make changes. I want things to get better. When it came time for me to decide to pursue the political side and run for sheriff, that was my motivation. That also made making the transition easier for me. It also helped that I wasn't raised a cop and only saw what law enforcement agencies were like and how they operated. I was able to employ elements of my business background. I also understood the distinction between being a cop and being a politician or elected official.

I also think that even early on in my career, when I first went to work on an Indian reservation as a member of the police force on the Salt River Pima-Maricopa Indian Community, my maturity gave me an advantage.

I'm like any guy. I love competition. I love pushing myself. I loved mixed martial arts at the time. And working on the reservation required physicality. Part of the job was getting physical with people who weren't following the law and who were putting up resistance. But the key was to make sure that you kept yourself in check. You had to be aware of your temperature and other markers. Going in, I wasn't sure if I had that skill. I wondered if my temper would get the best of me. Am I going to be able to feather the throttle when I need to? I found out that I was. I was good at it.

I understood when it was time to be tough and when I needed to be kind. And guys go after adrenaline opportunities. We seek out that rush. But being older, that was just a small part of it. I needed to be doing something with more of a purpose. When 9/11 happened, I was thirty, and I spent every night on the internet trying to figure out how I could make it work to enlist. I'd thought about it when I was younger but got talked out of it. Now here I was—married with five kids—and I couldn't do that. But that hunger to do something that would make a difference never went away. Most jobs are about making money, but there's very little *purpose* to society involved. I didn't say benefit, but purpose. This is a job with a high purpose for society. Plus, I can go out and challenge myself and get those shots of adrenaline.

Also, I'm Mormon, and our church teaches us to find purpose in our life, to push yourself to constantly be better—learn more, be better spiritually, be better mentally, be better intellectually. Become your best. Cherish and preserve freedom and patriotism. This job enabled me to do all those things.

And when you go to work, you realize that it isn't all about enforcing laws. It's about helping people in some of the worst times in their lives. When I started this job, I felt like I was back on my mission in Argentina. The objective there was to be of service too. It takes a different form than it does as a cop, but it was great preparation for what I do now.

About three years ago, a reporter wanted to do a ride-along with me. I was fine with that. Five minutes into it, we got a hot tone. We shoot over to the first call. It's a motorcycle accident. We're the first on the scene. It's clear that a driver turned in front of him, and he slammed into the side of the car. Another deputy shows up right after us. I'm over the motorcyclist, a young kid, doing chest compressions on him. After a while, I swap out with another deputy so he can take over with CPR. I grab the victim's ID. He's twenty, a kid, and I see that he lives nearby. He's also a member of the church we attend.

I can tell that this kid isn't likely to survive. I say to the ride-along reporter that we need to go. I want to get to that family before they get a call later. I want to tell them about the accident and that their child is being taken to the hospital. We arrive at the address. The father comes out and he's very friendly, but I have to cut him short and say that I'm not here for a good purpose. I tell him what happened, and his demeanor changes immediately. He goes to get his wife. She's crying and she gives me a big hug. I had dated her in high school and she says, through her tears, "Oh Mark," and that's all she can get out. I'm glad that I can be there to console the family instead of someone they don't know at all. I spend some time with them at the hospital too. Then the rest of the night goes on and I take four or five more calls with the reporter still with me.

At the end of the ride along, the guy says to me, "Man, I didn't realize that you guys spend so much of your time helping people." I tell him that's mostly what we do. We don't always deal with bad guys. That's what people want to see or read about, but it's a small percentage of how we spend our time—people with domestic issues, mental health issues, drug issues. The reporter said to me, "You guys get such a bad rap, but you're hustling all night to help people."

It took that ride along for him to get it. People don't realize that we go out on first aid calls to assist in saving a life. With domestic violence situations, we're trying to save a marriage. Or we're trying to keep people from going to jail. And studies show that the average police officer will experience between forty-seven and a hundred traumas in their career. Compare that to the two to four the typical citizen endures. Police are drowning in trauma. Early on, you get sold on the idea of becoming a cop with a pitch that tells you that you'll be able to drive cars fast, you'll be able to shoot guns. They don't lead with the idea that you'll be helping people with their issues.

That's a problem because when they face those problems on a regular basis, a number of them will say, "This isn't what I signed up for."

I'm not saying it's fun to help people solve problems. You have to. But when leadership makes bad decisions about how to handle that, the problems start to bother us more and more. Society, frankly, is catering to mental health issues. It's not trying to fix it. It's guiding us into more problems. Some say that you can be whoever you want to be—a man, a woman, somewhere in between. That's just ludicrous. And the effects of that kind of permissiveness and "do whatever makes you feel good" kind of thinking spills over into the community. Now we face people who think they can do whatever they want and without consequences. In law enforcement, we're picking up the pieces of this mental health approach that everyone has bought into, pouring gas onto a fire, and then we have to show up and deal with the consequences of that perspective.

In regard to leadership failings, most of the chiefs and other people in positions of leadership in law enforcement in the US have pulled back the reins, and will tell you they want to hold the cops accountable, as opposed to saying that society has some problems. Nobody's here to tell you that every cop is good, but I'm not going to sit and let anyone make the narrative that cops are the problem. That's just not the truth. The majority of the cops are very good people who go out and do a job every day and are forced to make split-second decisions, even when the optics of what we do don't look good.

We need leaders who stand up for their officers and stand in front of them when the bullets are flying. If you do that, you make it easier for those under your leadership to find the less desirable parts of the job a little bit more doable.

★★★★★★★★★★★★★★★★★★★★★★★★★★★★★★

> I hope most people who read this book feel a lot like that reporter. I want this book to be a literary ride-along, like Mark shared with him. Reading Mark's thoughts on that night, to go from pounding on a young man's chest desperate to keep him alive, to immediately hugging his mother and consoling her. The thoughts and emotions Mark processed and the reporter witnessed are "all in a day's work" for most first responders, especially those who work in their hometowns and communities. As a society we have little insight and even less appreciation for what this work demands and what service these professionals provide. But if we're going to learn it, understand, and appreciate it, we need to be enlightened by stern and consistent leaders who demand we see the good in their work.
>
> Because of Mark's views on good leadership, he decided to take the next step and run for the office of sheriff of Pinal County, Arizona, in 2016. To that point, he'd enjoyed a stellar career. He earned valedictorian of his academy class, was recognized as Rookie of the Year of the Salt River Pima-Maricopa Indian Community police department, and later was awarded Officer and Detective of the Year. He served on the SWAT team and as a detective he worked in cooperation with federal and local authorities to bring down dangerous gangs in the region. He was clearly well qualified as a law enforcement officer and had risen in rank, but Mark felt that wasn't enough to earn a community's or an agency's trust as a leader.

★★★★★★★★★★★★★★★★★★★★★★★★★★★★★★

AN OUTSIDE-THE-BOX APPROACH TO LEADERSHIP

In law enforcement, the system isn't set up to identify the best leaders. It's set up so that an individual can check all the boxes and earn a promotion. You go through a process of filing the paperwork, taking a test, and if you fail, take those same steps again. Once you pass the exam, you are promoted. Keep doing that and you are promoted to a supervisory level. That doesn't necessarily mean that you're a good leader, just that you're a good test taker.

Now, I'm a brother in blue. I'm with my fellow officers and supervisors all the way, no matter who you are. But I was looking for someone who would inspire me to walk into battle with him and be excited about the opportunity to do that. I believed, generally, that law enforcement needed better leadership. Under the Obama administration, I didn't like the direction the country was going. I saw the rule of law being undermined. There was a concerted effort to tear apart the fabric of trust that people had with the police. Body cameras were being introduced. All that did in my view was erode the trust that people had in the police and in the court system. It used to be if, as an officer, you went to court to testify, what you said happened was accepted as the true account. We saw that change. It wasn't about your testimony. It wasn't about the written report you filed. It was more about going to the video, and what was seen was subject to interpretation.

I saw society shifting, and people undermining the rule of law, while cops were being made the bad guys and the criminals the good guys. We've seen that shift continue even more in the last four years. I didn't want to be a guy who just complained about the lack of leadership and this shift. I wanted to do something about those problems, and so I ran for sheriff.

When you get elected by the people, you earn a mandate to do what you told them you were going to do during the run-up to the election. I'm different from a lot of elected officials, including other sheriff's candidates. They don't want to make direct statements of intent and execution. They also don't want to share what's taking place once they are in office. I took the opposite approach because I believe that's the best way to go about doing the job that I told people I'd do, and they put their trust in me. I wanted to push out a realistic narrative, the truth of what we do. We're cops. This is how we handle this aspect of law enforcement, this scenario. We showcased it on live TV. We showed how we handle traffic stops. We showed how you get processed into jail. We did this on social media as well.

Only one agency in this country has more Facebook followers than Pinal County—the FBI. We surpassed the New York City Police Department. I believe that no other law enforcement agency in the country has more YouTube followers than this sheriff's office. I didn't work toward those goals for self-aggrandizement. I did it for very legitimate reasons. First, in order to hire the best people, you have to market your agency. Second, you have to let your community know what we're doing. And using social media, I was presented with an actual gauge that I could read to tell me how the community felt about certain measures we were taking or programs we were instituting. You get the results and you fine-tune.

As I see it, a lot of agencies live in a bubble. They don't know if the community supports them or to what degree they do or don't. When you look at what happened in Minnesota with the George Floyd killing, that incident exposes an agency that doesn't have a relationship with the community. So the people immediately become the enemy. In our community, in response to all the negativity about police and policing, we said, no, that's not the case here. Our cops are good. Our sheriff's office is great. The community communicates with us all the time. The same was true with our approach to Covid. We weren't going to impose the masking or vaccine mandate. We weren't going to enforce the stay-at-home orders. The community and the police got together to figure out a way that worked.

I brought a business sensibility to being a sheriff and running the sheriff's office. We provide a service. The people are the consumer of that service. We tell them how we do things and what we offer, and they tell us whether they like it or not. And you have to understand the needs of those consumers. What do they want? By and large, they want to feel safe. They want you to protect them. They want someone to stand up for them. They don't want all kinds of other BS. And the thing is, that's what your officers and other sheriff's department employees want. That's what the citizens want. That's the beauty of being a sheriff. You're beholden only to the people.

We were fortunate to get our feet underneath us before some of the really tough times hit, when the shift in attitude toward policing really turned negative. One of the first things I did when taking office in January 2017 was hire a public information officer. I wanted to get someone who wasn't a cop. I hired a person with a media background. It was easier to teach her what she couldn't say, what might jeopardize an investigation or a prosecution, than it was to teach a cop how to be creative. So we took the creativity piece, and I used my own creativity to make it happen and to make it real. That helped set us apart from how things had been done in the past.

I'm a believer in bank accounts. You have one of those with the trust from your community. You have to make deposits in that account every day. You build it up little by little. You rarely are able to make a large deposit, like saving a dog from drowning. That would be huge trust deposit in the community. You've got that account building, but inevitably there will be some withdrawal, usually due to something big happening. An officer has to shoot someone. There's a bad interaction between an officer and a citizen, and that gets pushed out all over on social media. If you don't have money in your account to cover that withdrawal, you're in trouble.

So, I drum it into everyone in the sheriff's office that we need those daily deposits. We have to get money coming into the account every day. That way, when the bills come, we can pay them. That's how we found success. We set the narrative. We didn't let the media set it for us. That allows our guys to go out and do their jobs. And the results have been very positive. We've reduced the crime index numbers in our community year after year since I've been sheriff. And we've been able to do that as an agency and a community at a time when crime is on the rise in so many places. And we're able to hire good people, even though we're at the tail end of the Phoenix Valley and other agencies are able to offer higher pay, because we can say that it's a great place to work and we get results.

One initiative that I had us set up is a youth redirection program. I wanted to establish a more positive relationship with young offenders. How you treat them early on will influence how they act later. If they have a bad view of police officers because of an early experience, then they're going to carry that forward into adulthood. In four years, they're going to be voters. In eight years, they're the people whose calls your officers will be responding to. You want to have a positive relationship with them. For that reason, I don't like seeing kids get charged with petty crimes like marijuana possession.

To set up this sheriff's youth redirection program, I went to our county attorney and shared what I was proposing—which, in the early days, wasn't anything more than that I wanted to do something with youth to achieve those goals. I wanted some latitude to develop it and needed some input from that attorney. I also wanted to see what was being done in this area in other places. I traveled to California and made a few other visits. After thinking about it and doing that fact-finding, a clearer vision emerged.

I didn't want it to be a diversion program where the case goes to the county attorney and then, instead of being handled through the court it gets taken care of outside of that process. Rather, a deputy has an interaction with a kid. We look at that as an interception rather than letting the county attorney handle it. We don't arrest them and get the courts involved in some cases. Instead the deputy fills out a referral form so that offender can be a part of the program they have to attend.

We conduct four classes: goal setting, planning, digital media/social media/sexting, and drugs and alcohol. We also do a career night. As much as possible, with great frequency, I'm running those classes. Along with the classes, these kids are required to do eight hours of community service. We also offer horse therapy, which is very, very effective with these kids. Two Saturdays, they will be introduced to working with or caring for horses.

More than two hundred kids have gone through the program, and maybe five of them offended again. That's a very good measure of success, but that doesn't mean those kids don't have any problems anymore. We kept them out of the legal system. We kept them out of jail. That saves the county a lot of money, while also saving a lot of those kids. Many of them can join the military because we wiped their records clean once they finished our program.

We do other things, like offer tattoo removal programs in the jail. If you've got gang tattoos and you go looking for work, you've reduced your chances of getting hired. Or those gang affiliations send a negative signal to others in the community. We created a "Humvee"—a pod, a housing unit within our jail for military veterans. We provided them with camouflage jumpsuits so when they make an appearance in court, the judges will know who they're dealing with. We do a reentry program for adults so the transition out of the legal system is better. This has been tremendously effective. We've had four thousand inmates who've been positively affected. We've cut down some of the jail days they have to do—a cost saving and a reward for being a part of the program. We've seen a reduction in recidivism since instituting that program and others. It isn't just them going through the program and checking a box. We're getting them help and they're not coming back to jail.

I worked a lot of gang crimes and intervention before becoming sheriff. So I dealt with a lot of kids. I could see what precipitated them joining gangs. I wondered what we could do to break that cycle of gang involvement and criminality and ending up in the system. One thing about kids that they have in common is that they need people in their life. They want someone's time and attention. But what's better than that is positive interactions. What we could do as a sheriff's office is provide the positive. These are kids at a very vulnerable part of their lives, and we could get to them and move them in a positive direction so that we won't have to deal with them again down the line.

★★★★★★★★★★★★★★★★★★★★★★★★★★★★★★★★★★★★★

> There has been a lot of talk about "law and order" in American politics as of late. I fully support both those things in our society. But a tagline isn't a policy and a politician is rarely the one getting his or her hands dirty to affect this law and order they promise.
>
> That's where Mark, as an elected sheriff, stands out. He might not have had to, but for him "getting his hands dirty" meant being creative and energetic in actually solving problems, not just dealing with the aftermath. His forward thinking required a look back. Back to where hardened criminals were before they committed these crimes, what influenced them and how could they have been pushed in a better direction. He went upstream and intercepted them before they became a full liability. Such a tactic requires a tremendous amount of empathy and work.

But it also requires a leader who knows when someone is truly rotten to the core, or if they might be someone worth investing in. Mark's age and experiences prior to joining law enforcement helped him find creative and effective ways to connect or "market" to his community. Along with his business experience being brought to bear in his approach to running the sheriff's office, Mark's time working on the reservation is different from what most cops have done. Policing in that environment is much different than in other American communities.

★★★★★★★★★★★★★★★★★★★★★★★★★★★★★★★★★★

POLICING THE RESERVATION

First off, Indian reservations are sovereign nations. That means they have their own laws that may differ from state laws. They also have their own court system. So there is a lot to learn—state and tribal laws and procedures and the rest. In the case of a native offender, a tribal member, their case could also be handled in federal court. If it was determined to be handled in federal court, things got problematic sometimes because of the Indian Crimes Act. Within that act, only thirteen criminal acts are enumerated.

Let's say we arrested a tribal member on a charge of doing a drive-by shooting, our hand was forced. That individual couldn't be prosecuted in federal court because the crime they committed wasn't among the thirteen crimes listed in the Indian Crimes Act. That meant, because they were a tribal member, their case would be prosecuted in tribal court. The maximum sentence for that offense in tribal court was one year. I can't think of many instances where that maximum sentence was imposed. What that meant for us was that offenders were cycling through the system quickly. We were dealing with repeat offenders all the time.

Some individuals had as many as thirty DUI citations but they still had their license. The tribal court doesn't report those offenses to the state. That's one example of the challenges you face. But there are positive elements to working on the reservation. For example, they tend to be smaller and less populated. You can see measurable results of your work making a difference. I became a gang and drug detective. I worked drive-by shootings on the reservation.

For some perspective: The reservation is located in the center of a triangle of large cities—Mesa, Tempe, and Scottsdale. There's a total population among them of more than a million people. From 2000 to 2007, those local agencies handled 58 drive-by shootings. On the reservation, with 5,000 people residing there, we handled 200 drive-bys in that same period. In 2009 alone, things got even worse with 76 drive-bys. These were gang-related activities.

We had to do something. We worked with agents from the Bureau of Alcohol, Tobacco, Firearms and Explosives (ATF). The FBI was less involved but contributed to this joint effort, along with a couple of local agencies. The data let us know that one gang was responsible for nearly every one of these shootings. We decided to put a case together. We came up with a list of thirty individuals involved in these crimes. We narrowed that down to fifteen to pursue as part of a RICO investigation.

★★★★★★★★★★★★★★★★★★★★★★★★★★★★★★★★★

If you watch shows like *Sons of Anarchy* or *The Sopranos* like I do, you're probably familiar with the term RICO. For those of you who prefer rom-coms or sitcoms, RICO stands for Racketeer Influenced and Corrupt Organizations Act. Most often you hear the term and think of the mafia and maybe Tony

> Soprano. What the reservation police department and the other agencies did was essentially to classify a gang as what it is—a corrupt organization. Without going too far into the weeds, making it a RICO involved some technicalities of policy and procedures. More important, it became a federal case. It also allowed the leaders of the gang to be prosecuted even if they didn't commit the individual crimes. They could be prosecuted for setting in motion the illegal activities that others carried out. Think of it as being able to cut off the head of the snake.

★★★★★★★★★★★★★★★★★★★★★★★★★★★★★★★

USING RICO TO HELP THE RESERVATION

We were able to make the case. The federal prosecutors were able to get the convictions. The leaders and others went to jail. The number of drive-by shootings was 76 in 2009. We put that case together. In 2010, there were 10. In 2011, there were zero.

So, we made a big impact. That's not just a measurable difference, but a big difference. And the community felt safer. That was cool. That felt good. I got to know a lot of people in that community and enjoyed working on the reservation because of that. We were being effective, and to start a career wanting to be of service and being of service was rewarding and instructive.

It also helped that I was being well paid. I had a take-home car. But still, I wanted to be able to impact more lives. I saw the leadership vacuum in other parts of the area. I knew that if I ever wanted to make an even bigger difference, I'd probably have to run for sheriff. That meant leaving the reservation department and working for the county. In that large jurisdiction, I'd be exposed to more people. So I made the decision to leave. I left one of the best jobs as a gang and drug detective and joined Pinal County, taking a $25,000-a-year pay cut and becoming a deputy and not a detective. It all worked out, we made it work out, and in 2016, I did run for sheriff and won.

I only had about nine years of experience when I ran, but I packed a lot into those years. My philosophy was that the only way to get better at the job was to spend more time on the job. I had a sense of where I wanted to get to, and knew that at age thirty-four, I didn't have a lot of time to get there. So I took extra shifts all the time when working on the reservation. And I knew that was the place where I wanted to start out. I never even applied anywhere else. My family's roots in Arizona go way back. I have ancestors who came over on the *Mayflower*. Maybe more pertinent, my great-great-great-great-grandfather settled parts of Arizona. His name was Daniel Webster Jones and he had lived on the land that became a part of the reservation. So I had ties to the place and a genuine fondness for it. Something deep inside me was attracted to that place.

Most people don't know that the power of the sheriff is not given in the United States Constitution. But the Founding Fathers were well aware of the role that sheriffs played in the colonies and elsewhere in the world. Every state has a provision in its constitution regarding the powers of the sheriff. Those powers vary from state to state. Generally, out west, sheriff's offices are more full-service. Arizona is one of the states where the sheriff's powers are more expansive. We saw that here during Covid. I could override federal or state mandates. One negative consequence of that is that more and more people on the left, because they don't fully understand the concepts, believe that the constitutional sheriffs' movement is a negative thing. What they fail to understand is that it was always intended to be this way. Sheriffs function as a part of the coveted idea of separation of powers.

We didn't ask to be arbiters of bad laws that go against the Constitution of the United States. States can consistently enact more laws that are not really constitutional. And when they do, that forces us to act in accordance with the oath we swore in taking office—to defend the Constitution of the United States and, in my case, of the state of Arizona. But when a law violates the US Constitution, then who stops that in the interim while the courts adjudicate that matter? Somebody has to step in to say hold on, we can't allow this to happen. That's part of the role of being a sheriff. We can tell the government to pound sand if there's a law, an order, or a mandate that violates the Constitution we swore to defend. The Founding Fathers wanted the separation of powers among the three main branches, and as a sheriff, we get thrust in the middle of this. We're the people that have to defend the Constitution and people's rights under the law.

I always tell people, "My job is being the county sheriff. I'm not the government. I'm here to protect you from government overreach. That's my job."

★★★★★★★★★★★★★★★★★★★★★★★★★★★★★★★★★★★★★★★

Mark hits on a very important aspect of policing that many of us, and even some police officers, don't fully understand. He talks about the broad powers most sheriffs have in their office and duty. However, when you really process what he's pointing to with his job in respect to the Constitution, you see that what some may see as power, he sees as responsibility. Not just the obvious responsibility of working burglaries or murders, the evil acts we commit upon each other, but his responsibility to protect our rights, the ones given to us by God and guaranteed in the Constitution. He knows that the worst violator of those rights are our governments themselves. We need more leaders like Mark, not just in law enforcement, but in all forms of government. With great power comes great responsibility. With great leaders like Mark come great results.

★★★★★★★★★★★★★★★★★★★★★★★★★★★★★★★★★★★★★★★

ACKNOWLEDGMENTS

In recent years, and over the time this book was being written, our country has seen extraordinary natural disasters, tragic accidents, and acts of pure evil. As I have covered these horrific events and the loss of life and property they create, I know we often neglect to acknowledge the heroism displayed and trauma suffered by those who rush in to save lives and stay to clean up the mess.

The 2023 Covenant School shooting, as tragic as it was, featured heroism by the responding officers that will educate police departments all over the country for years to come. Hurricane Helene hit the southern United States in the fall of 2024. The devastation across the South but especially in Eastern Tennessee and Western North Carolina was unlike anything most of us have ever seen. Search-and-rescue teams flocked to the Appalachian Mountains to help however they could.

Shortly after the clock turned to a new year on January 1, 2025, a terrorist used his truck to mow down more than a dozen people in New Orleans. Police officers ran to the scene, engaged the driver, and neutralized him within minutes, saving countless other lives as we later discovered the terrorist was armed with guns and bombs.

Just a week later the entire world was shocked to watch entire towns in Southern California go up in flames as the Santa Ana winds hit hurricane speeds creating a true firestorm. Hotshot-trained firefighters from all over the state and even Canada responded, and even as more fires grew, they never stopped. They evacuated communities and fought fires in one of the most densely populated areas of the country and never stopped until the fires were fully contained weeks later. And just a few weeks after that, at the end of January 2025, a military helicopter collided with a passenger jet at Ronald Reagan Washington National Airport in Washington, DC. Although all souls aboard both aircraft were eventually confirmed lost, dive-and-rescue teams braved freezing cold water all through the night trying to find any survivors, then they proceeded to take the same risks for the next several days to recover as many bodies as they could for their mourning families.

This event was followed just a day later by a passenger jet soaring into the ground in Philadelphia just outside a neighborhood. As it exploded upon impact, jet fuel set a fire that spread as far as a half mile. Again, in what looked like a war zone, police and firefighters responded immediately, taking heroic action to save lives. These events were tragic, some inexplicable and even evil. But they are what our first responders train for and/or respond to each and every day of their careers.

We remember how we felt watching it unfold on TV; now let's acknowledge how they must feel running into the abyss of events like these to risk their own lives to save the lives of strangers.

Johnny Joey Jones
February 2025

ABOUT THE AUTHOR

JOHNNY JOEY JONES provides political commentary across all Fox News media platforms, including Fox News Channel and Fox Business Network, also serving as a fill-in host for many of the most popular shows. Additionally, he hosts several series and specials on the network's digital streaming service, Fox Nation. A Marine Corps veteran who reached the rank of staff sergeant, Jones suffered a life-changing injury in Afghanistan, resulting in the loss of both his legs above the knee. Since his recovery, he has dedicated himself to improving the lives of all veterans and their families.